澳门基金会

※

北京外国语大学

中国海外汉学研究中心

※

大象出版社

合作项目

卫三畏文集
Selected Works of Samuel Wells Williams

佩里日本远征随行记
（1853–1854）

The Journal of S. Wells Williams:
Expedition to Japan With Commodore Perry
（1853–1854）

宫泽真一　（主）
周 国 强　（副）　转写、整理
宫泽文雄　（副）

大象出版社

图书在版编目(CIP)数据

佩里日本远征随行记(1853—1854):英文/[美]卫三畏著.
[日]宫泽真一等转写、整理.—郑州:大象出版社,2014.1
ISBN 978-7-5347-7963-3

Ⅰ.①佩… Ⅱ.①卫… ②宫… Ⅲ.①日记—作品集—日本—近代—英文 Ⅳ.①TS805

中国版本图书馆 CIP 数据核字(2014)第 011228 号

卫三畏文集
佩里日本远征随行记(1853—1854)
宫泽真一(主)
周 国 强(副) 转写、整理
宫泽文雄(副)

出 版 人 王刘纯
责任编辑 李光洁
责任校对 钟 骄

出版发行 大象出版社(郑州市开元路 16 号 邮政编码 450044)
发行科 0371-63863551 总编室 0371-63863572
网 址 www.daxiang.cn
印 刷 河南省瑞光印务股份有限公司
经 销 各地新华书店经销
开 本 787×1092 1/16
印 张 28.25
字 数 411 千字
版 次 2014 年 2 月第 1 版 2014 年 2 月第 1 次印刷
定 价 165.00 元
若发现印、装质量问题,影响阅读,请与承印厂联系调换。
印厂地址 郑州市二环支路 35 号
邮政编码 450012 电话 (0371)63956290

主编序言

陶德民　张西平　吴志良

2008年8月8日，在北京奥运会开幕的当天，新落成的美国驻华使馆举行了剪彩仪式。美国国务院历史文献办公室为此发行了纪念图册，题为《共同走过的日子——美中交往两百年》，其中把1833年抵达广州的卫三畏(Samuel Wells Williams，1812—1884)称扬为美国“来华传教第一人”，虽然他并非是第一个来到中国的美国传教士。图册对卫三畏在促进美中两国人民的相互了解方面所作的贡献作了这样的概括和评价：“传教士成为介绍中国社会与文化的重要信息来源，因为他们与大部分来华经商的外国人不一样，这些传教士学习了中文。例如，美国传教士卫三畏就会说流利的广东话和日语。他曾参与编辑英文期刊《中国丛报》，供西方传教士及时了解中国的最新动态，方便在美国的读者了解中国人的生活。卫三畏还编辑出版了《汉英拼音字典》（即《汉英韵府》——序者）和分为上下两卷的历史巨著《中国总论》。时至今日，他依然被公认为对19世纪的中国生活认识得最为精透的观察家。”

图册中也提到“蒲安臣于1862年成为第一个常驻北京的美国代表，在促进中国的国际关系上扮演了更为积极的角色。蒲安臣迅速结识了一批清政府的主要改革派官员，其中就包括咸丰皇帝的弟弟，设立了中国第一个外务部（应为总理各国事务衙门—序者）的恭亲王奕訢，”以及他1867年辞去驻华公使后作为清朝使节率团访问欧美，与当时的美国国务卿签订《蒲安臣条约》一事。“该条约扩展了中美两国的往来领域。条约规定，两国人民享有在彼此国家游历、居住的互惠权，并可入对方官学；中国领事在美国口岸享有完全的外交权利；鼓励华工移民美国；美国政府支持中国的领土主权。”其实，蒲安臣的卓有成效的外交活动及其维护正义的外交主张，与当时担任使馆秘书兼翻译的卫三畏的鼎力相助是分不开的。卫三畏的儿子卫斐列后来倾注极大热忱为蒲安臣撰写传记，其重要原因之一也在于此。

今天，当我们出版这部文集以纪念卫三畏诞辰200周年，并重新探讨其经历和成就之际，有必要思考这样一个问题，即卫三畏一生所体现的诸多志趣和倾向之中，什么是值得我们加以特别关注的呢？虽然这是一个“仁者见仁，智者见智”的问题，多数人恐怕都不会对以下见解持有异议，即卫三畏在对华态度上所发生的“从教化到对话”的某种转变可以说是他留给后世的最重要的精神遗产之一。而这恰恰可以从耶鲁大学校园里留存至今的两件文物上得到印证。

第一件文物是该校公共大食堂(Commons Dining Hall)墙上的卫三畏肖像画，它其实是设在纽约的美国圣经协会所藏同一油画的一个复本。众所周知，卫三畏的在华生涯大抵分为两个阶段，作为传教士的前22年(1833—1855)和作为外交官的后20年(1856—1876)。在前22年的最后一段时间里，他还作为首席翻译官随同佩里将军的舰队两次远征日本，在改变日本锁国政策的外交谈判过程中发挥了重要作用。佩里将军对其语言天赋和交涉手腕的赏识，无疑是卫三畏转而步入为期20年的外交生涯的一个契机。1877年辞职回国后不久，他受聘为耶鲁大学的中国语言文学讲座教授。1881年春被遴选为美国圣经协会会长后，他便提出希望有才华的画家制作自己的肖像画，并将其献给圣经协会。结果，由耶鲁大学艺术学院的韦约翰教授制作的油画所呈现的是这样一个形象：等身大的卫三畏站立在放有纸、笔和一个中国式花瓶的桌子边上，目视前方，手持1858年《中美天津条约》的文本。

很显然，这样的构图反映了卫三畏本人的意思：他一直把基督教传教自由的条款得以列入1858年《中美天津条约》看作自己毕生的最大成就。这是因为当时负责谈判的美国公使列卫廉对是否要在条约中加入该条款并不经意，只是在卫三畏的坚持之下才获得了最后的成功。此一事实，不仅可见于耶鲁大学斯德龄纪念图书馆附属档案馆所藏卫三畏的亲笔日记，也可以由当时协助他进行谈判工作的丁韪良牧师后来所写的《花甲忆记》得到证实："那份现在成为条约荣耀之处的条款是卫三畏博士提出的。"在预定签约的6月18日即"决定命运的那天早上，卫三畏博士告诉我他一夜未眠，一直在考虑这份宽容条款。现在他想到了一种新的形式，可能会被对方接受。他写了下来，我建议我们应当马上坐轿子直接奔赴中方官邸解决这个问题。……中方代表接见了我们，他们当中的负责人稍作修改，就接受了卫三畏博士的措辞"(详见陶德民《从卫三畏档案看1858年中美之间的基督教弛禁交涉——写在〈基督教传教士传记丛书〉问世之际》)。1876年夏天，美国国务院接受卫三畏的辞呈，在正式解职通知中对他赞扬有加："您对中国人的性格与习惯的熟悉，对该民族及其政府愿望与需求的了解，对汉语的精通，以及您对基督教与文明进步事业的贡献，都使您有充分的理由自豪。您无与伦比的中文字典与有关中国的诸多著作已为您赢得科学与文学领域内相应的崇高地位。更为重要的是，宗教界不会忘记，尤其多亏了您，我们与中国订立的条约中才得以加入自由传教这一条。"而卫三畏则在复信中坦承，这一嘉许是"最最让他感动与满意的"(见卫斐列著，顾钧、江莉译《卫三畏生平及书信——一位美国来华传教士的心路历程》)。顺便提到，在1858年6月19日，亦即《中美天津条约》签订的第二天，美国驻日总领事哈里斯便与德川幕府的代表签订了《美日修好通商条约》。卫三畏闻讯后，又不失时机地向美国的新教诸团体提出派遣牧师赴日传教的建议，并得到了切实的响应。所以，对美国基督教会的东亚传教事业来说，卫三畏确实是一位有功之臣。

那么，卫三畏是基于一种什么样的信念而企图以基督教义来教化东亚的所谓异教徒的呢？试举一个例子。1853年7月16日，在佩里将军初访日本，成功向日方递交美国总统国书一周之后，卫三畏曾从停泊在江户湾的军舰上给他在土耳其传教的弟弟W．F．威廉斯写了一封长信，其中指出："佩里告诉日本官员，他将在次年率领一支更大的舰队以求得到他们对所提要求的回答，即所有前来访问或遇难流落日本海岸的美国人应得到善待，美国汽船在一个日本港口得到煤炭以及有关物资的补给。这些是我们花费巨大开支和派出强大舰队到日本水域的表面上的理由，而真正的理由是为了提高我们民族的名誉和得到称扬我们自己的材料。在这些理由的背后并通过这些理由，存有上帝的目的，即将福音传达给所有国家，并将神旨和责任送达这个至今为止只是在拙劣地模仿耶稣之真的民族。我十分确信，东亚各民族的锁国政策决非根据上帝的善意安排，其政府必须在恐怖和强制之下将之改变，尔后其人民或可自由。朝鲜人、中国人、琉球人和日本人必须承认这唯一活着的和真实的上帝，他们的锁国之墙必将为我们所拆除，而我们西方的太平洋沿岸城市正开始派出船队前往大洋的彼岸。”从卫三畏对佩里访日使命的解读来看，他固然赞成打开东亚各国的大门，但更重视把上帝的福音传入各国人们的心扉，期盼由此引发一系列的变革。因为他认定上帝所代表的真善美的品格和力量是美国等基督教国家所拥有的绝对优势和无比强势的源泉（详见陶德民《十九世纪中叶美国对日人权外交的启示——写在日本开国150周年之际》）。

然而，卫三畏在以后长期驻节北京的岁月里，逐步加深了对中国的悠久历史和深厚文化传统的了解，使他在对华态度上开始发生“从教化到对话”的某种转变。而这一点可以耶鲁校园里的另一件艺术杰作来作为象征，那就是在卫三畏于1884年2月16日去世两周之后，在举行其丧礼的巴特尔教堂（Battell Chapel）东北角中心位置安装的一块非常别致的彩色纪念玻璃。玻璃的中央写有《论语》中的七个汉字：“敏则有功，公则说。”（意思大体为，勤勉能积成事功，公道则众人心悦诚服。）玻璃的下方用英文记载了卫三畏的名字、生卒年份以及他一生所扮演的角色与身份，即“传教士、学者、外交官、耶鲁学院中文教授”。这块玻璃由纽约的路易斯·第法尼公司（Louis C. Tiffany & Co.）负责制作，其别出心裁的设计显然是出于卫三畏生前的授意。

卫三畏选用孔子的话作为自己一生的概括，可谓恰如其分。勤勉能积成事功的道理，可以从他孜孜不倦编纂而成的《中国丛报》、《中国总论》以及《汉英韵府》等多种刊物和名著得到了解。公道则众人心悦诚服的道理，则充分体现在了他辞职回国之际所受到的各界朋友的赞扬。他在信中告诉他的妻子，“临行前威妥玛于周六、西华于周一为我举办了两次特别宴会，我见到了北京的所有名流，并受到了他们的盛赞。然后，似乎是为了区别于外交聚会，丁韪良、艾约瑟、怀定先后请我吃饭。我告诉过你，我在丁韪良家见到了所有的传教士，那真是一次令人难忘的聚会”。“接连不断的登门送别者中有三四位是总理衙门的官员，京城中最位高权重的大官也来为我送行。除了亲王之外，我都是在办公室里会见他们，彼此依依惜别。九位官员每人赠我一柄折扇作为纪念，这充分显示了中国人的友善”。不仅北京的新教传教士们发给他一封送别信以示友谊与敬意，上海的同仁也发给他一封相似的信，恰如其分地总结了他在中国的生活：“您长期担任美国使馆秘书、翻译，九次代理公使的职务，这些工作给了您许多重要的机遇，使您得以把知识、经验用于为中国人造福、为您自己的国家谋利，尤其是为基督教在中国的传播效力。对您工作中表现出的高度责任感，我们不胜钦佩。”对这些热情洋溢的赞美之辞，卫三畏则表示自己的所作所为只不过是在执行上帝的旨意，“我在传教过程中与同伴们相处融洽、身体健康、工作愉快，为此我要虔诚地赞美造物主”。

在某种意义上说，卫三畏选用孔子的话来概括自己的一生，也显示了他对中国文化传统的敬重之意。而这种敬重之意，可以由1879年他为反对加州的排华风潮而撰写的《中国移民》中的一节得到佐证：“当加州的法庭想用立法来反对中国人时，它将中国人等同于印第安人的简单态度是颇为古怪的。生理学家查尔斯·匹克林将中国人和印第安人归为蒙古的成员，但加州的最高法院却认为‘印第安族包括汉族和蒙古族’。这样在概念错误的同时，它还支持了一种错误的观点。它把现存最古老国度的臣民和一个从未超越部落关系的种族相提并论；把这样一个民族——它的文学早于《诗篇》和《出埃及记》，并且是用一种如果法官本人肯于学习就不会叫作印第安语的语言写就，而它的读者超过了其他任何民族的作品——与最高的写作成就仅是一些图画和牛皮上的符号的人群混为一谈；把勤奋、谨慎、技艺、学识、发明等所有品质和全部保障人类生命和财产安全的物品等同于猎人和游牧民族的本能和习惯。它诋毁了一个教会我们如何制作瓷器、丝绸、火药，给予我们指南针，展示给我们茶叶的用处，启迪我们采用考试选拔官员的制度的民族；把它和一个轻视劳动，没有艺术、学校、贸易的种族归为同类，后者的一部分现在还混迹于加州人中间，满足于以挖草根过活。”虽然卫三畏在赞扬中国人的同时，也表达了他对印第安人的歧视态度，他的观点还是得到当时耶鲁学院全体教授的同情，他们纷纷签名支持他起草的请愿书，呼吁海斯总统否决1879年的反移民法，其中包含这样的警告：如果我们的政府首先改变条约的规定，中国政府也可以根据国际法取消治外法权，致使美国在华公民失去领事的保护。此外，1878年中国北方发生大饥荒时，身处太平洋彼岸的卫三畏利用他对美国在华传教士和商社的影响，不遗余力地推动救灾。他还试图劝说国会归还1859年中方赔款余额的一部分来帮助缓解灾区的困境。而早在1860年，他就曾建议国务院用这笔赔款余额在中国建立一所美中学院，以培养中方的翻译、通商和外交人才。这个建议得到了新上任的总统林肯的同意，只因为当时的国会未予批准而搁浅。卫三畏这个未能实现的建议完全可以称之为后来退回庚款吸引中国学生留美的先驱性方案。

卫三畏还对亲眼目睹的中国同时代的缓慢而确凿的变化做了一定的正面评价。1883年7月，即辞世的半年多之前，卫三畏在儿子的帮助下终于完成了《中国总论》这一巨著的修订工作，在增补版的序言中他写下了如下字句：“我在1833年到达广州时，和另外两位美国人作为‘番鬼’向行商经官正式报告，在他的监护之下才得以生存。1874年，我作为美国驻北京公使馆秘书，跟随艾忭敏公使觐见同治皇帝，美国使节站在与‘天子’完全平等的地位呈上国书。由于一生之中的这两次遭遇，并念念不忘在思想和道德上的重大进步是使一个孤傲的政府从强加于人的姿态改而听从他人所必需的，毫不足怪的是，我确信汉人的子孙有着伟大的未来，不过，唯有纯正基督教的发展才是能使这一成长过程中的各个冲突因素免于互相摧残的充分条件。无论如何，这个国家已经度过被动时期，这是肯定无疑的。中国不可能再安于懒散隔绝—像过去那样，以过于自负的态度俯视其他国家，就像面对她所无需劳神的星星一样。”

如果说，17世纪的礼仪之争最后是以天主教士在康熙禁令之下被逐出中国告终的，那么，这场可谓19世纪的礼仪之争则是以“天朝上国”在第二次鸦片战争之后所做出的巨大让步来结局的，包括废除以夷相称、公使驻京、平身觐见皇帝而不必行三跪九叩之礼。卫三畏的字里行间所透露的，既有在礼仪上争得与最古老"中央王国"之"天子"平等地位的自豪，又有从其所信奉的上帝的立场针砭这个王国"懒散隔绝"和"过于自负"的高傲。联系到上述1853年他给在土耳其传教的弟弟W. F. 威廉斯的长信，可见这时的卫三畏虽然在其对华态度上开始了"从教化到对话"的转变，但并未把"异教徒"放在完全对等的位置上，特别是在道德和精神的层面上。在这个意义上说，自以为占据着道德精神的制高点，只有基督教可以拯救中国的卫三畏也难免"过于自负"的诟病。这就是笔者在前文中提到其转变时，始终不忘加上"某种"一词来予以限定的缘故。

尽管如此，卫三畏在序言中宣称，修订版"坚持初版序言中所述的观点——为中国人民及其文明洗刷掉古怪的、模糊不清的可笑印象"。卫三畏相信，在传教"事业取得成功的基础上，中国作为一个民族，在道义和政治两方面，将会得到拯救。这一成功有可能和人民的需要同步前进。他们将会变得适应于自己着手处理问题，并且和外国文明以多种活动形式结合起来。不久，将会引进铁路、电报和制造业，随之而来的是中国人民中成百万地受到启发，无论在宗教、政治和家庭生活的每一方面"。这种力图呈现中国文明及民族的美好品质和衷心期待中国在对外交往中日益开放进步的善意态度，自然是应该予以充分肯定的。

从卫三畏的例子不难联想到，身处所谓"轴心文明"（主要以几大世界性宗教作为标识）的人们往往患有"老子天下第一"的自恋症，而以"居高临下"的态度傲视其他人，难以免俗和根治。反观我们自己，历代国人的"华夷之辨"又何尝不是如此？而时至今日，仍不乏"宅兹中国"的自尊和输出"天下体系"的自负。但也不必悲观，因为这个五彩缤纷的世界本来就是由形形色色、各有偏好的人们构成的。他们之间有相辅相成的关系，也有相反相成的因素。虽然各派宗教和文化的代表自以为有"替上帝传教"或"替天行道"的使命，谁都无法垄断真理和剥夺他人的信仰，也不能指望按照自己的面貌来彻底改造他人。其出于宗教或政治动机的种种努力的后果，客观上会推动各国和各民族之间的交往和融合，并推动人类文明的进步和升华，则是毫无疑义的。从这一点来看，卫三畏一生的变化过程所显示的方向对我们建构今日地球村的和平共处规则还是不乏启迪作用的，那就是要逐渐学会把"老子化胡"的心态改为"相敬如宾"的心理，对异民族、异文化和异文明采取更加宽容、尊重和善于调适的态度，以便用文明对话来取代文明对抗，从而把亨廷顿认为不可避免的"文明冲突"消弭于萌芽状态和无形之间。

献　呈

To Timothy E. Head,
Going Native in Hawaii

Contents

Preface

While visiting the Archive of the Yale University Library to prepare for publication in China of the correspondence of Samuel Wells Williams (22 September 1812 – February 16 1884)[1], I found the three bounded volumes of his diaries. One is what you have now in your hands, his daily journal of the Perry's Expedition to Japan for 1853 and 1854[2]. The old veteran Commodore of the U.S. Navy, Matthew Galbraith Perry (1794 – 1858) succeeded in putting an end to the isolation policy of the Shogunate Government by his peaceful, though threatening with men-of-war, means. Japan now opened their long-closed door to the foreigners and their commerce, which important event caused as great an impact to all the Japanese as the Opium War had done to the Chinese people. 島崎藤村 (Shimazaki Tohson: 1872 – 1943), modern poet and novelist, made an impressive sketch of such a far-reaching shock and influence of the coming of the Perry's black ships in one of his great novels called *Dawn*.

S. Wells Williams liked correspondence with his family, friends and colleagues. What we sometimes call his "Japan Journal" in short are actually not a diary, but it is the collection of numbered sheets of letter-paper which were written, folded and sent by person or mail to his wife, as private letters first of all. After his death, his only son,

1.《美国耶鲁大学图书馆藏卫三畏未刊往来书信集》（全 23 册）广西师范大学出版社，2012 年 12 月。

2. Yale MS547/Series II. Samuel Wells Williams Papers/Box 30. *Journal: Trip to Japan with Commodore Perry 1853-1854.*

Frederick Wells Williams edited the private letters of "Japan Journal" for the sake of publication in 1910 for the Asiatic Society in Japan. We can see that Frederick did edit his father's writings. For one thing, he cut quite a lot of whole pages and passages especially referring to the family affairs and personal religious life of his father, which amounts to something nearly 50 pages in all. He sometimes thought it better to change the words and expressions of his father into his, paraphrasing or editing, as well. We can notice a few misprints which apparently were overlooked in proof-reading by Frederick.

My principle of preparing a transcription text out of whatever augraph papers, such as letters or diaries, of S. Wells Williams, is mere faithfulness to his original writing. The working procedure is something like careful and patient reading of his autograph writing to start with, then preserve or represent it in every way possible to our modern computer usuage.

It is hard to say who made the original sheets of letter-paper into a bounded volume, perhaps by S. Wells Williams himself, dating back to 1860s to 1870s, or by his son Frederick dating back to 1900. We can understand that the bounded volume was better than a collection of numbered and folded sheets of letter-paper, when it came to printing in Japan. On the other hand, a bounded volume is likely to suffer much more damage, for a spine and thread is often broken by rough hands of compositors. Actually the preservation condition of our volume is extremely bad indeed, tord pages, spine broken, etc. Since it is apparently hard to have the damaged volume microfilmed, the only choice left to me for reading and typing it was to spend one summer at New Haven to read and transcribe it from the original document, another summer to make a digital camera photo copy of the volume, and the third summer for my son, 宫泽文雄 (Miyazawa Fumio: 东北学院大学大学院博士课程), one of the assistant editors, to spend two weeks at the Library to get my own digital photo copy re-shot or corrected. He pointed out so many parts of my first draft trascription text were not faithful representations of the original text. We fear that it costs so much time and cost in restoration that perhaps for some time to come we may find it difficult to prepare any facsimile reprint edition of the "Japan Journal" volume for publication.

宫泽真一 (北京语言大学汉学研究所客座研究员), chief editor, is responsible for the transcription text of this volume, while 周国强 (长崎县立大学国际文化学部), the other assistant editor, helped him with proof-reading and notes.

I tried my best to be faithful and exact to the original writing of S. Wells Williams. Notes in square brackets are those of editors, while a few notes in round brackets are those of S. Wells Williams himself. Frederick, his son, added a number of scribblings in red pencils on the original text, and also he changed the way of speaking of his father etc. I ignored them all of the later changes or mistyping on the part of Frederick, purposing only to be as faithful a transcriber to his father as possible.

It is a great pleasure to dedicate this humble work to my old university teacher, Timothy E. Head, who wrote an interesting book about Hawaii, remembering his interest in the history of the Bonin Islands when I met him first time in 明治学院 half a century ago.

宫泽真一 (Shinichi Miyazawa), September 2013, Beijing.

1853

On the 9th of April, 1853, I received a request from Commodore Perry to accompany him to Japan as interpreter, he wishing to have me ready by the 21st on which day he intended to sail. On his reaching Canton, I had an interview with him, and learned that he had made no application to the Secretaries [of ABCFM] at Boston respecting assistance of this sort, nor informed them of his intentions; he said that this never occurred to him, for he had repeatedly heard in the U.S. that I wished to join the expedition, and would be ready on his arrival in China to leave. Dr. Bridgman [Elijah C.Bridgman:1801~1861;an American missionary to China, ABCFM][1] was with me at this interview, and we spoke of various topics connected with the enterprise taken in hand to improve the intercourse with Japan, from which we inferred that this first visit this year was intended to chiefly ascertain the temper of the Japanese in respect to the propositions which would be submitted to them. At any rate no hostilities were determined on, except, indeed, to repel an attack or actual aggression, for many vessels of the squadron had not reached China yet, and he wished to make an experimental visit first. He added that he had refused to employ Von Siebold [Philipp Franz Balthasar von Siebold:1796~1866] as interpreter because he wished to keep the place for me, —doubtless a compliment to me, but not very wise in him, so far as efficient intercourse [~~to~~] with the Japanese went.

In conclusion, I told him that unless I could get some person to take charge of my printing-office, I could not possibly leave Canton. At the next meeting of the mission, held April 20th, it was concluded that Mr. Bonney [Samuel William Bonney:1815~1864;an American missionary to China, ABCFM] leave his station at New Town and find somebody to take the house if possible, and take charge of my printing-office while I was absent; he intended, if possible, to get Mr. Beach or Mr. Cox, if not both, to occupy the house, but in this he failed.

I went to see Com. Perry the next day, and told him that I would go with him till October, and could not be ready to leave before the 5th to 10th of May, in consequence of the various matters necessary to be attended to. It was recommended to him to get a lithographic press in order to assist in promulgating the wishes of the American people & let the people know what we had come for; to this he agreed, & I purchased an iron press of Mr. Lucas for $120, which I hope will be a good outlay. I stipulated too, that I should not be called on to work on the Sabbath & should have comfortable accommodations on board ship. Moreover, I stated to the Commodore that I had never learned much more than to speak with ignorant Japanese sailors, who were unable to read even their own books, and practice in even this imperfect medium had been suspended for nearly nine years, during which time I had had no one to talk with; he therefore must not expect great proficiency in me, but I would do the best I could. In my own mind, I was almost decided not to go at any rate on account of the little knowledge I had of Japanese literature and speech, and am now sure that I have been rightly persuaded by friends to go. It is strange to me how attention has been directed to me as the interlocutor & interpreter for the commander of the Japanese expedition, not only from people hereabouts but from the United States; while we are here, speculations as to the propriety of leaving Canton in this capacity, a letter comes from Plattsburgh, desirous Sarah to come home with the children, for that her friends had heard that I was to be absent two years to act as interpreter in Japan. I certainly have not sought the place, nor did I expect more than to be consulted as to the best mode of filling it.

On Monday evening, we had a pleasant meeting at my house at monthly-concert, where all were present; the expedition to Japan was particularly commended to the prayers of all interested in the furtherance of the Gospel. Dr. Hobson [Benjamin Hobson:a British medical missionary;London Missonary Society][2] read an extract from the "Chronicle" respecting the change in the policy of the Queen of Madagascar, showing that the persecution suffered by the Christians there for many years was to cease, & full liberty likely to be granted them thro' the powerful influence of the heir-apparent: & the son of the prime-minister, both of whom had become favorable towards Christianity. Mr. French [John Booth French:1822~1857;an American missionary;Presbyterian Board] remarked that this association at this meeting of Japan & Madagascar, reminded him of the last monthly-concert he attended in America, at which they were both brought to notice, and particularly prayed for; & the happy change in the last made him hope that a favorable result might follow this attempt on the latter. May God in his infinite mercy grant that this expedition be a means of advancing the latter-day glory, when the heathen shall be the people of Christ, and then I shall be rejoiced that I have gone with it. At any rate, a beginning must be made in breaking down the seclusion of the Japanese, and I hope this attempt will be blessed to that end.

All my preparations being made, & my teacher appearing with his baggage, I left Canton, May 6th, in the steamer for Macao, to join the "Saratoga", and sail to Lewchew. I was greatly annoyed in getting aboard to find that the lithographic press & materials were not there; but it came down by fast boat before sailing, for I found that

Capt. Walker would not sail till Tuesday, in consequence of the want of bread, and Mr. Bonney forwarded it on Friday evening. I spent a few days at Macao very pleasantly, and on the forenoon of Tuesday, the 10th of May, I set foot on board ship, & sailed on the evening of the 11th, nearly sixteen years since I left in the Morrison for the same region. Of my fellow-passengers there, Mr. King [Charles W. King:b. 1808 or 1809;an American merchant in Canton][3], Mr. Gutzlaff [Karl Friedlich August Gutzlaff:1803~1851;Netherland Missionary Society; married twice to English ladies], Capt. Ingersoll [Captain D. Ingersoll:an Amerian sea captain of the ship *Morrison*], and three of the Japanese, are dead. It was mentioned by Com. Perry that I had a strong inducement to go with him from having been in that ship, as the inhospitable treatment received by the "Morrison" was to form one of the reclamations of the present visit. How vast a change has happened in the politics of China since that cruise, in opening her principal ports and commencing a freer intercourse with her people; when we returned in Aug. 1837, not a port on the Chinese coast was accessible, and nothing known of their capabilities.

Wednesday, May 11th

We were to sail today, but an untoward event this morning delayed the ship. One of the crews had been locked up in the cell yesterday in consequence of his outrageous conduct when under the influence of spirits, of which he evidently had taken a large quantity. He was an active seaman but quite ungovernable while possessed with rum, and his conduct merited punishment. This morning he was found dead in his chair inside of the cell, greatly to the surprise of all, for he had been visited only a few minutes before, when he refused his breakfast. An examination into the circumstances showed that he had taken a bottle of brandy out of the spirit-room while at work there, & lest he should be detected, he had drank it nearly all off within a few minutes (half an hour), making excuses to get away from the room to take a draught out of the bottle. He soon became ungovernable, & was shut up in a cell, where his noisy bawling & singing disturbed all the watch during the night, and showed that he was still unfit to be liberated. He died without a struggle, probably of some interference in the action of the heart. The corpse was taken ashore in the afternoon by a boat's crew, having been encoffined and carried around the ship before the assembled crew, the marines presenting arms & others uncovering, as the body passed by. So he died, this James Welsh, as a fool dieth; for no "drunkard can inherit the kingdom of heaven". Yet the grog bucket is daily brought on deck, and all who please take a cupful of the mixture, which tends to strengthen the appetite, and confirm every one in habits of intemperance. It is unfair to them, for the crew could easily be shipped without its promise; and it is unfair to the officers, for the source of trouble is continued, while they are forbidden to whip those who may offend.

Saturday, May 14th

We are now fairly on the way to Lewchew, and are likely to have a head wind all the way up the Formosa channel. Pedra Branca was passed this morning, and the coast was just visible in the distance; the water is smooth, the sky overcast, the wind dead ahead, and the ship deeply laden. Ashing, the pilot, left us yesterday, with last words to friends near & far, which he is to deliver to Mr. Desilver. The officers, besides giving him his liberal wages for the use of his boat, $40 a month, and $20 for his pilotage, also presented him with presents of food & money, & he went off with crackers firing.

I am hardly able to compose my thoughts yet to study or read to much purpose, for the novelty of the place, the number of people about, and the motion, tend to distract me; I have begun to look over some phrases in Japanese, which Giusabaru wrote many years ago. The more I think of it, the less satisfaction do I find in the prospect before me; it was none of my own seeking, however, and I can only do my best.

The news from Shanghai of the insurgents being in full possession of Nanking, which they were fortifying with Chinkiáng & Yangchan fú, is trifling compared with the reports brought by Mr. Meadows [Thomas Taylor Meadows:a British interpreter;author of *The Chinese and Their Rebellions*,London,1856][4] of their camp being governed entirely on the purest Christian principles, that they are Christians in all respects, and take the Bible for their rule of action, observe the sabbath and preach a pure monotheism to all those around them. If half we hear proves to be true, truly a new day is dawning on China.

Sunday, May 15th

There were no services held today of a public nature; no work was done, and generally the ship was quiet, men engaged in reading. It is a bad arrangement which leaves the holding of public services so completely in the hands of the commander, tho' as we have no national church, it is not easy to say what rules could be laid down on this subject.

I have been thinking, in respect to the supposed successful result of this expedition, how soon the merchants in China would try the sale of opium along the coasts of Japan, and do all they could to induce the people of the country to consume it. How to avert such a sad result is beyond my sagacity, for no laws can reach the appetites of a people, no scruples will embarrass the seller in placing the temptation before them, and their moral principles are not likely to stand against a seductive luxury. This view would be more saddening, if one did not remember that the mixture of good & evil in this world is necessary for the development of the probationary plan on which this world is governed; and that God overrules all, and will make the wrath & avarice of man to praise him at last.

May 20th

On Tuesday, a strong wind arose from the north, causing the ship to pitch and roll about in the chopped sea caused by the same wind, and making everybody uncomfortable, and me seasick. I was soon unable to do anything but lie as still as the jerking of the vessel would allow, and passed a most uncomfortable day. The violence & direction of the wind induced the captain to change his course about noon, & steer for the Bashees. Next morning the wind had ceased, leaving us under the lee of the Pecadore Is.: and about 3 P.M. the breeze sprung up from the eastward, and completely heading us off on our course thro' the Bashee past as it had up the Formosa channel. Yesterday, it was nearly calm all day, but this morning a light, 2 knot breese sprung up. For three days, the men have been drilled by one of marines, and marched up & down the quarter-deck; this is to make them expert at the musket, and ready for an emergency. Among other things, they have been firing at a mark hung up at the yardarm, which most of them hit — it being a board painted like a man, and not a difficult target.

Friday, May 20th

I have been looking over the Japanese phrases I once wrote out with Giusaboro, but they do not easily recur to mind. I have forgotten almost all the phrases I once had at my tongue's end, and am afraid that nine years' cessation from using the language has obliterated most of it from my memory.

The men caught a dolphin the other day, a species of Coryphene, but I was too seasick to do more than go & see it. It was a yellowish green, with three dorsals & pectorals, the former reaching the whole length of the back. There were many in the water, and their bodies glancing thro' the spray, at the bow, presented a very pretty appearance.

Tuesday, May 24th

The weather & wind were pleasant & favorable till yesterday morning, carrying us forward at a rapid rate along the eastern shores of Formosa; we had a distant sight of the south end of Formosa and of Botel Tobago sima, too far to see anything more than their outline, however; no other land has since been seen. Yesterday morning, the moon was full, & a change of weather took place, the wind coming from the NE. with rain & squalls, and making everything & everybody uncomfortable. We are S.E. of the Madjïco simah group, and find a N.W. current setting us off to leeward, which is somewhat unexpected. Perhaps this current is formed by the wind blowing down the coast, & meeting the streams which debouch into the Yellow Sea, is driven off into the Pacific between Formosa & Lewchew.

Such motion disorders one who is yet unused to it, & I find it almost impossible to attend to anything satisfactorily. Old Sieh [謝] lies abed most of the time, & seems to be getting weak & heady from the motion & confinement; he is old, and that indisposes him to exertion, besides the weakness which he feels from the disease of his opium or tobacco. I begin to be almost afraid he will not prove of much service to me, but I hope I shall be able to get him recruited by a visit on shore at Napa. — I have been aboard ship now a fortnight and a greater change can hardly have been passed over me than compare the life I have had all the spring at Canton, with this tossing, queasy and confined life in the "Saratoga". I suppose I shall be comforted for all discomfort by being told that "it will [do] you good"; but I shall be pleased to have it do me no hurt.

Nautical notes of trip to Napa. [There is a chart and minute notes, half a page, which were kept by SWW on board ship, beginning on May 12th and ending on May 26th; they are now not transcribed here in this volume.]

Thursday, May 26th

We made land yesterday afternoon, and not wishing to get in too near, stood off to SE. with a light breeze; but when we drew towards it again at sunrise, we knew not the land, as it did not agree with any view laid down on the charts, and it was not until we had drawn up along its western side, opening one island after another, that we ascertained that the ship was westward of the Amakirrima Is., to which we had been drifted by a strong westerly current during the night. We had passed by so as to open the main island, when we saw the two steamers coming up on the N.W., the "Susquehanna" taking the lead, & the "Mississippi" a mile or so astern. We gradually wore up, having a scant wind, and when Napa opened were far to the N.W., and to leeward, with small prospect of getting in to an anchor. After the steamers went in, a shift of wind enabled us to lay in from the N.W., and by sunset we reached the place, and dropped anchor within a cable of where Ingersoll placed the "Morrison" almost sixteen years ago (July 11th, 1837), and found a patch of 10 ft. which I am glad to see that Beechy's chart has called very properly "Ingersoll's patch". The feelings arising in one's mind at returning here, and remembering the party and their hopes, with whom I was then connected, are of a mixed character; the residence of Dr. Bettelheim[5] & his family is a great advance on the position of things then, & this is the entering wedge of more extended operations of others.

May 27th

At 9 o'clock. Capt. Walker & I went aboard the "Susquehanna", where we found Mr. Jones, Bittinger & Bettleheim engaged to breakfast with Com. Perry. We discussed various things at the table, & after breakfast, Bettleheim made known to me his ideas of things as he had already spoken of them to the commodore. His position & opportunities for intercourse have greatly improved during the last few months, & many restrictions have been removed; he has visited the north part of the island, and the people are not ordered away as they used to be. About 10 o'clock, Lt. Contee & I went ashore with him to see the local magistrate（地方官）of Napa, & tell him the reason why the presents he sent to the "Susqa" were refused. We landed near Capstan Pt. and went up to Bettleheim's house, where we waited while the messenger went to announce our visit to the "mayor of Napa", as Bettleheim calls him. Meanwhile we talked with B. & his family; he has three children, one of them born here, & lives in a pretty comfortable way, at least it looked so in the bright sunshine. In an $1^1/_2$ hour, it was announced that he was approaching the <u>kung kwan</u>, having gone to the other hall near the jetty. We saw, on entering the [place] a considerable group of well dressed people, & the old mayor came forward & bowed. He was a venerable look[ing] man of 62, dressed in yellow robes. We took seats, and [we] informed him thro' an interpreter, that we had come on [the] most friendly grounds, and wished to have amicable intercourse; that we declined the presents, for the reasons that none were allowed to be received by our laws, & we wished to buy our supplies; we also wished to see 總理官 or Regent, on board ship tomorrow, and would there tell him what our wishes were, and how long we were to remain here probably. He could not say whether the Regent would come off, but made no opposition to the request; it was also intimated that a house would be wanted ashore for a hospital. This hint caused some stir among the retinue, but all feeling was repressed. During the interview pipes, tea & refreshments were handed about, and every civility was offered us. The groups forming around us from time to time were very picturesque, silently looking on or else whispering among themselves, they walked around or squatted down, there being no other chairs besides ours. The room was matted, & open to the air, inclosed in a yard defined by coral walls; the whole forming a pleasant-enough place for conferences. In $^3/_4$ hour we left, and returned to the boat, the mayor accompanying us to the gate-way, and the silent crowd still looking on. The street is one of the largest in the town, & many groups were stationed here and there at the entrances of houses; coral walls defined the grounds around each dwelling, and gave rather a dull appearance to the avenue; tho' it was lively enough now with people.

We reported progress to the commodore, & at dinner with him I met Lts. Hunter & Randolph. A room is preparing for me on the [saffrail] of the steamer, in which I shall be comfortable in warm weather. During the day, no one but the party sent has been ashore, but the boats have visited the reefs & picked up shells & other things.

May 28th

About 10 o'clock, the commodore sent a boat for me & my teacher, but on reaching the flagship, I was surprised to receive a letter from his hands written by Bettleheim, couched in the strangest style of intreaty and advice respecting the conduct of the expected visit of the Regent to the flagship, and concluding with the hope that the natives would not come near the ship, which I myself more than thought would be the upshot of it, for no promise could be given by the persons I saw yesterday. It was about the oldest mélange I ever read from Bettleheim, whom the commodore had sent for, & erelong reached the ship. He soon was all in motion, & it was about concluded that if the Regent came off, Com. Perry should not see him. However, about $12^1/_2$ o'clock, he was announced, and Capt. Buchanan took him into his cabin; he was accompanied by the interpreter I saw yesterday, and several other officials, some with yellow & their attendants with red caps, while the regent himself had a striped paper cap, all of them of a square shape, like a blacksmith's paper cap. Only the chief man sent his card 琉球國中山府尚大謨. A few formal compliments were passed, and Capt. Buchanan rose to conduct him about the ship, which took about an hour or so, and rather exhausted the old gentleman. The whole party showed considerable interest in the vessel and its inhabitants, which indeed must have amazed them if they have human ideas. The Commodore, after reflection, concluded to receive them in his cabin, and tho' I had for a little while been swayed by what Bettleheim had said, I was not sorry that he saw them, for the party came at his invitation to see him, & why not receive them? They had brought a trifling present of two cakes & two jars of spirits, which were to be accepted, & it was meet to thank him. All came into the cabin, and having been seated, it was told them that their visit was received as a mark of kind respect, and that the American nation entertained the most amicable feelings towards Lewchew, and that the present visit was to open further intercourse with it. The proximity of the two countries across the Pacific ocean was stated, and something said of California and its gold. Refreshments were handed around, and all partook; wine & cake being articles intelligible to all, and the Regent's attendants brought in pipes. The Commodore taking one with him. He seemed half stupefied, at times, but it was probably amazement at his novel position, for he was frequently speaking to the interpreter. A motion to rise induced Perry to say that he should be ready to return the visit on the 6th prox° at the capital in Shui 首里 and thank him for his civilities. Excuses were offered that it was far, that the king was sick, that the visit was a mere form, & the presents contemptible & beneath notice. However, it was stated that propriety required him (Perry) to return the visit, & he should not fail. The decorum of these islanders on board, and their subdued way of looking about, did them credit. A barometer was shown them, a revolving pistol too, and the rudder was moved to & fro, the tiller ropes having attracted their attention. Nothing was here said respecting a house on shore, and all conversation with them on general topics was very slow & almost impractical from their anxiety and the tedious line of communication. The Regent rose and left the cabin; & when on deck, Capt. Buchanan took him into his own room, there to take a glass of wine, and would have kept him a while but he seemed to be desirous of going. The band played several airs which pleased them all, & the marines drawn up in order, the huge guns & large balls on deck, were objects of great interest. The party left after a visit of about 2 hours; [~~and~~] a few of them seemed to enjoy it, but such a melancholy set of faces, fixed, grave, and sad, as if going to execution, was hardly ever before seen on board the Susquehannah. Bettleheim talked a good deal, and his way of making signs, and motioning with his face, was very much disliked, and wrongly interpreted. I hardly know what to think of the man, for he whisks about in his opinion like a weathercock, and after the Regent had gone, said it was the best thing whh. could have been done, to see the Commander, tho' his letter of four pages was to urge the contrary.

After dinner, we went ashore to B.'s house, where Mr. Barry made out a list of provisions, to be given in to the flagship tomorrow. Major Leilen also went to see a level place where he could drill his marines, & from that we visited the tombs of some foreigners buried on shore. I also left Sieh on shore at B.'s house to recruit a little.

May 29th

It rained all day, & I remained aboard the "Saratoga", unable to go to service in the "Mississippi", where it was thought there would be no preaching. Bettleheim sent back Sieh, in the boat which brought the provisions to the "Susq.", & wrote a letter to Com. Perry about interpreters. In the evening I took Sieh to the flagship, & gave him in charge to [~~Asmo~~] Achin, Perry's servant, by whom he will be cared for.

May 30th

The drizzling rain of yesterday cleared off with a pleasant sky, and enabled the "Caprice" to get in to her anchorage this morning. She has not had very pleasant weather, & leaks in her deck. I saw Mr. Maury soon after his arrival, & was glad to see him looking so well. The commodore sent an order on board ship today by Bettleheim, for Mr. Gouldsboró [*sic*], Mr. Harris & myself to accompany him ashore, & get a house for the transaction of business. This order certainly carries with it a decided tone, and I am not sure how we shall manage, in carrying it out. However, we went off, Mr. Madigan, Lowrie & Stockton going with us. Mr. Bettleheim took us along the street beyond the bridge at Tumai, the same which I remembered to have passed by when we came ashore in this direction from the Morrison, the morning after we anchored. About half a mile from the landing-place, he came to a public-hall, which we entered after the door had been opened by persons crawling over the wall. A messenger was straightway dispatched for the mayor of Napa, and after waiting an hour or more, the interpreter alone came with two constables or lower officials, to whom we made known the commodore's application for a place on shore convenient to remain at and see about matters connected with provisioning the ships. The request seemed strange to them, and they said it was a better way for us to give lists of things wanted, and the articles would be brought off. We rejoined, that it seemed but decorous, after the Regent's visit to the "Susquehanna", that a convenient place should be rented on shore, from whence men could be sent to each ship with what was wanted. The interpreter said there was no place fit for us, there was none vacant, there was no need of such a place, & that the house we were in was a schoolroom, as indeed we saw it was partly used for some such purpose. We talked to & fro in this way a long time, Adjirashi, the interpreter, at last going off to see the mayor, while we remained for his answer. It was then concluded, that as it was expedient to make a right out of our might, so we had better, if we wished to get a house at all, keep possession of this; two of us were therefore sent back to get bedding & our dinners, while I went to report at headquarters, where in truth I got but little satisfaction or even approbation. On returning ashore, the messenger had not yet come back, but while Mr. Goldsboro's & others were on their way to the boat, they met him, and returned to the hall where many native officers were still tarrying. His answer was to the effect, and he could only still protest against our occupying the dwelling, notwithstanding he saw the bedding and other preparations we had made for remaining. It was a struggle between weakness & right and power & wrong, for a more highhanded piece of aggression has not been committed by any one. I was ashamed at having been a party to such a procedure, and pitied these poor defenseless islanders, who could only say no. No one was incommoded by the act, indeed, for no one lived there; but perhaps the towns people of Tumaï felt it all the more keenly, and I pitied them heartily.

Mr. Stockton & I were soon left alone with our three Chinese, for after Mr. Goldsboró left the house, the natives officers retired, and we made oursleves as comfortable as we could on the thick mats which covered the floor; but the fleas & musketoes would not permit us to sleep, and the Chinese walked about all night. A large company of Lewchewans occupied the other rooms, and kept watch over us, if the insects let them do so, for they were very still, except an occasional hum. The dawn showed that it was time to rise, & I was glad to get into the fresh air, & terminate my first night in Lewchew, the unwilling agent, in so doing, of violence and wrong.

May 31st

Mr. Lowrie came erelong to receive [~~me~~] us, and when I reached the ship, I heard Mr. Goldsboró say that Perry approved of all we had done, & was decided to keep the house, and was going to send two or three invalids there to keep possession. During the forenoon, he (Gold$^{o.}$) went to the house to see about arranging for the comfort of the invalids, & while he was there, the mayor of Napa came in with the interpreter, Idjïrashi, and had a long talk with him respecting holding it. This man has had considerable instruction from Mr. Bett$^{m.}$ & during the talk he made out to converse on many topics, referring to place, in China, countries in Europe, America, &c. He said he had heard of Washington being a good man, but he thought Washington would not have done so. A written protest was handed in to make known to the commodore the desires of the authorities in regard to the house, couched in respectful terms, in which, however, were two or three misstatements.

The general feeling on the whole, among the people, seems to be more & more favorable to us, and they are learning a few things gradually. The constant presence of officers & men ashore familiarizes them with us, and the crowds of idle people are as large as ever. Boat-loads of visitors throng the flagship, and the crew are glad to show them this & that.

Wednesday, June 1st

Went ashore with Purser Barry, when we learned that the authorities will not acknowledge our presence in the house we have taken, and provisions must be forwarded thro' their purveyors, who will receive lists from Bettm. only. It is surprising what a degree of quiet resistance, an organized government like this can offer to violence, without any overt act of violence, without giving any excuse for the wrong by doing the like themselves. They feel their weakness & have no intention probably of resisting by force; but the complete sway they have over the common people enables them to wield what power they have to the best advantage. I need cite only one fact: wishing to go to the "Saratoga", I hailed a boat which had just left the steamer, & went in her; as she left to go ashore, I threw a bowl of cash into the boat, but it was with much trouble brought aboard, tho' it could have been divided among them without one being more interested in keeping it secret than another.

We went up to the house to bring away a sedan-chair for Perry's use, and found all quiet; Capts. Buchanan & Adams were there, and had brought Perry's answer to the petition to Goldstorô sent in yesterday. On the way back to the boat, one of the Chinese carrying it stopped to look at a market by the roadside, and his contemptuous look at the beggarly assortment of leaves, pottery, fuel, and eatables, was not more amusing than the gaping wonder of the women & people at his gigantic (6ft. 2 in.) height compared with their Lilliputian size. I never before saw such a lot of hags together as in this market.

After having put Perry's answer into Chinese, old Sieh went ashore by mistake, and [~~by~~] in his stupid way left behind, & had some trouble in getting the natives to take him aboard. I supposed he would have gone to Bettleheim's house instead; he does not recruit much, & I am afraid will die.

The "Caprice" goes tomorrow, & many are sending their clothes over to Shanghai to be washed, as there is little prospect of getting it done here. The letter-bag takes Bettleheim's first letter sent off for 11 months, besides $800 sent over to put in the Bank there to his credit — his "own sweat & blood", he says. He says that he has not been able to come to any explicit understanding with the rulers or people as to the price of the provisions he consumes; they brings food and he lays down money, & no accounts are drawn out; he [sets] what they bring, they take away what he lays down.

Thursday,June 2nd

I moved my baggage over to the "Susquehanna" before breakfast, and spent some time in getting to rights there. After copying out the reply to the mayor's petition, I went ashore with Betthelheim with it, and after waiting a short time, we were informed that he was waiting to receive it at the town-hall. (Com. Perry had sent a cake to Mrs. B., & the children were eager to get a taste.) On reaching the town-hall, we were much surprised to see the Regent there, and a feast spread out on 5 tables, with a large crowd of officials in attendance, the whole indicating considerable expectation for somebody. I went in & handed the paper to the mayor, who was seated at the table, and said that he would look at it by & by, & showed no idea of opening it there. We bowed to the Regent, and soon learned that the party was waiting for the Commodore, who I suspect had no idea of the matter. It seems that they either did not, or would not, understand the declining of this feast, to which they had invited Perry on Tuesday, who could only reply verbally at the time their card came off to the ship. I knew not that any written invitation had been given, for Perry had never said a word on the matter; & therefore I could only say to Idjïrashi that I knew nothing of it, nor whether the Commodore was coming, except that he was busy & had not intimated his intentions. The matter was miserably managed, anyhow, for a written invitation was probably sent, for Achin told me that the authorities had invited Perry, & I think a written card would not be neglected. If he had a paper in Chinese he did not understand, why did he not find out what was told him? A written refusal was the least the authorities could expect: the feast was proposed by them doubtless as a means of avoiding a meeting at Shui; & this refusal gives them a handle, & not having had a written refusal, a longer handle, to take exception at granting that interview.

I reported the matter to the Commodore, who said that as he had had only a verbal invitation, he gave only a verbal refusal. The impression of a show of some sort was very general among the people, for there must have been 5 or

600 people in the streets, probably waiting for the guests. About 3 o'clock, Adjïrashi & others brought a portion of the dinner on board the flag-ship, and if words could be received as denoting real feelings, they certainly learned the real reason for declining it. They said that as Perry could not come to the dinner, they had brought it off for him; and a pretty show they had made of it. The whole was taken away by the officers & men, & the natives went back, probably rather mortified at their reception, for nothing was offered them while on board, nor even a chair.

The "Caprice" sailed for Shanghai about 2 o'clock, & I just succeeded in getting my letters on board of her. In the evening, a theatrical performance was acted by the flagship's crew; one of the pieces was Box & Cox, with singing, proem, & other longer exhibitions, the whole being very creditable. I was told to their thespian skill, for it was my first attendance at such a place & performance. All the officers were spectators.

June 3rd

My quarters on the "Susque.'s" taffrail are likely to prove very commodious when completed; just now I am at rather odd ends. D$^{r.}$ Bettelheim wrote a letter to the Commodore in his usual singular fashion, (calling him "father," & [~~willing~~] desirous to obey his orders, and talking of "glorious mission," and the flagship a "throne," and Perry an "autocrat" whose glance should be law to the natives,) yet finding fault with everything which has been done, chiefly, as far as we can learn, because he was not consulted. Yet when he read Adams' reply in Perry's cabin yesterday he called it "excellent," & approved of it all. The man does not seem to know his own mind for a day, but evidently wishes to be consulted about everything, & have his advice followed. He is not at all backward in sending or begging for things, while he, Jew-like, puts his money in the bank. However, this must be added, that he cannot spend much money here for his family, even if he wished, for he is not allowed to buy at will; & this sum may be the surplus of his salary. This P.M. he visited the flagship to report the result of the regent's colloquy with him, and bro't a petition from the Regent to the effect that the Queen Dowager was exceedingly ill, having never recovered from the alarm caused by the visit of Captain Shadwell in the "Sphynx" in Feb. / 52, and begged the Commodore to repair to the prince's hall, where a personal interview could be held. He also proposed an exchange of another house in place of the one now occupied, & mentioned a temple, as suitable for our use.

In reading such a document, one can hardly explain all its features by either Chinese or Japanese policy. The form of a petition (which is the constant style here towards foreigners) indicates a kind of servile feeling, which their consistent persistance in upholding what they call & hold to be law, rather denies; and their duplicity in these papers shows conscious weakness, which their complete control over their own people again contravenes. The oligarchy of the gentry tyrranize [*sic* / tyrannizes] over the people by means of moral suasion, which to have its present effect must have been long exerted and commenced in youth. The Chinese classics are regarded as the standard of morals, and certainly here show what a means of degrading in human mind they can be made, crushing all responsibility and paralyzing [~~all~~] the industry of the mass.

In the evening, our walks led out to the pier & by the junks, and no change seemed to have been made here since

1837. A score of junks lay in the harbor, some after the Chinese model, & some building of the Japanese fashion. A watering party of Japanese sailors passed by, but we saw none ashore, nor a large number in the junks. The market place for vegetables was full of people, & all the sellers were women, perhaps 600 of them, most of them remarkable for their long coarse hair and plain features. The police follow us everywhere, making no opposition, nor warning the people away, but yet acting as a check to intercourse. Two articles of interest are seen in the streets, & there are no shops of wares opened anywhere. There were not many buyers, and little alarm was manifest, tho' the women would always leave their baskets when we approached. The streets of dwellings are dull-looking by reason of the almost uniform dead wall in front of them, but these walls of coral are usually well built, and look as if they had stood many years. We tried to enter no houses, and saw few entrances so arranged that even the yard could be observed. The people occupy five times the space which Chinese do, but their comforts I suspect are not proportionate to the larger ground they occupy, though as a whole, they seem to be well fed. Their sober, downcast faces take away much from their looks, and repress all attempts to make one's self understood by talking to them.

June 4th

I was kept in the ship all day, preparing the presents, and drawing out the reply to the Regent's petition, telling him that he (Perry) must go to the palace, & if the other house suits this purpose, he will change to it. The old teacher was loth to take up his pencil, but we got it ready by 2 o'clock, and was just on the point of sending it in a boat, when to our general surprise, the Regent himself with his usual retinue came aboard. He was received by Capt. Buchanan in his cabin, and on being seated, handed another petition to him for the Commodore which was merely another request not to come to the palace, as the Queen dowager was very sick, and the Regent's house was the spot to repair to. He wished, but unavailingly, to see Perry, who would not appear. We declined taking their paper down to him, for after reading it we told them the answer was already contained in the answer now handed to him; this they deferred to open while on board. Capt. Buchanan offered them some drink so strong that they could not take it, for all I know it was clear brandy; he showed his unwilling consent to have them remain long in every action, & this was increased by Bettleheim appearing, who it seems had been invited off by the Regent, to facilitate intercourse. However, it was no use; they could not see the Commodore nor get any other answer than a reference to the paper handed them. It was a childish visit, and one hardly knows how to act towards such children, who must be, in a manner coerced for their own good. To talk about the principles of international law being applicable to such people is almost nonsensical; they must first be taught humanity and self-respect.

Before leaving they designated a man to accompany an officer to the other house they are willing to have us occupy, which proved to be the one formerly occupied by Forcade. Owing to the fresh breeze, Capt. Buchanan sent the Regent ashore in a cutter, and was glad to be rid of them. Bettleheim had a long talk with Perry; he is becoming more than ever disliked by everybody, & took an unlucky step in coming aboard today, when he was unwished.

I came across the Regent's invitation to dinner a day

or two ago, so that the contrements [sic / contretemps] might have been avoided if Perry had laid by the paper less carefully.

Mr. Jones & his party returned today, and gave a good report of his trip, and said there was much more to be discovered, & hoped another opportunity would be given of exploring the island towards the extreme north.

Sunday, June 5th

Mr. Bittinger preached today, & I dined with Mr. Jones in the other steamer. Few persons went ashore today; but a man-of-war is a miserable place to spend a Sabbath in, and a ramble on shore is worse. I learned more concerning the trip up the island, the conduct of the people and the officials, the extent of the cultivation, and the general character & productions, from Mr. Jones, than I had been able to do. He estimates the population at 100,000, more than 3/5 of whom live in the southern third.

June 6th

By half past nine, the party had reached the landing-place near Tumai, where it was formed in military order under the trees there, and started for Shui 守禮 about 10½ AM. The authorities had sent two guides, and provided 10 sedans & 4 horses, but in going up all preferred to walk, the day being very pleasant & agreeable, and they were told to follow after us. The guides went first, then came Bettelheim & I to see that they did not carry us to the wrong place. A party of sailors with two brass field-pieces under Mr. Bennett's command; a company of marines, the Mississippi's band, Com. Perry in a sedan chair, the coolies with the presents behind him, & a marine each side of the chair, the officers in undress uniforms, the Susquehanna's band marines, &c., amounting in all to over 200 men, made up the procession. As it passed up the well paved road, and wound through the defiles or turns in the ascent to Shui, it presented a beautiful appearance, such as no Lewchewan had ever before gazed on. The distance was about 3 miles, and nothing could have been more charming than some of the scenes which opened upon us as we advanced, temples, rice-fields, copses, houses, and walled inclosures, succeeding one another in pleasing exchange.

At the entrance to the capital, stands an honorary portal, bearing the inscription 中山, which means, I am told, the capital of the country. It was of fine proportions, the central gateway being 20 feet or so high, & the side ones 15 or so. Here commenced a level, macadamized road for the rubble paved one, and the walls on each side higher & solidly built.

Standing just beyond this portal, to my surprise stood the simpleton of a Regent with a large company of officers, and Idjïrashi came up to beg us to turn in at his yamun, which our guides were just about to do; Bettelheim too, wanted to parley with them, but I pulled him along, and said I would not speak with any of them. Thus we went on up to the palace-gate, a man running on ahead to open it, and our host trudging along in his slip-shod toe-thumb stockings by our side, putting himself by his silly conduct in a ridiculous position. I let him enter the gate a minute or two ahead, & then sent in the cards by Achín, for Perry now was nearly at the gate. One of the natives took me by the hand to beg me not to let the marines enter, & seemed vastly relieved by the assurance that they were not to enter. Near the gate was another honorary portal (*) like the other, with a different inscription 守禮之邦 Shui's domain. (* see Guillemard, vol. i., p. 54. for picture of this gateway.)

Going in, we passed through a second door into a yard, at the upper & raised part of which was a tripartite doorway leading into the palace yard, inscribed 奉神門 Door for receiving the gods; the authorities were all standing at a side-hall, the one in front being shut. When the principal persons were seated, a few formal questions were asked, tables were placed before us, (for the hall was perfectly bare of furniture) and tea & pipes introduced. The Regent & three Treasurers were seated in chairs opposite the Commodore & his captains. They had soon the list of presents in their hands, & presently arose to return thanks to the donor by a low bow. The Commodore then inquired after the health of the Prince & Queen Dowager, and offered the use of his physicians to assist in curing her. The Lewchewans seemed to have nothing to say to us, but rather to endure our presence; and Perry did not intend to introduce any topic. The hall, called the 高牖延薫 or High Inclusure for fragrant Festivities, was the same where Captain Shadwell delivered Lord Palmerston's letter, and like the rest of the establishment, very little used. No preparation had been made for us here, & the Regent begged us to stop a little while at his office on our return, which was agreed to; he had evidently made the preparation there.

The court-yards were paved in alternate strips of cut granite & sand, and was [*sic*] clean; the wood-work was painted when new, but now had begun to decay from exposure. The outer walls were built of stone much of it laid on the scarp of the hill, so that the outer look of the place was not unlike a fort, & was doubtless designed for some possible contingency of defence against insurgents; even now it could easily be garrisoned & fortified.

The Regent being evidently uneasy, his guests arose, & we were soon on the way to his quarters, Perry walking this distance with them. The people were not numerous in the broad way, & some saw the rattan laid over their backs when they incroached too near, in peeping thro' the bushes. This day was for the grandees, & the vulgar were not to intrude. The Regent had indeed gone to considerable trouble, there being some 15 tables spread with small sauces filled with cold viands, vegetables & drinks; and soon warm dishes were introduced. There were many yellow-capped officers standing about the room, but all the waiters had red caps, & most of them blue dresses of a pretty hue; the four high officers in their variegated caps, sat opposite Com. Perry, like so many Nestors, grave, silent & rather sad, — but nothing had spoiled their appetites, for they cleared most of the warm dishes. The Regent proposed to drink America; & Perry replied by the health of the Prince & Queen Dowager, & that our countries might always be at peace; all emptying their thimbles of cups each time. At the close, each party drank the other's good health; and we rose to leave before the 12 courses were all brought in, which Bettleheim said was a royal feast. There was no lighting up of the faces of the old men, & they were evidently wishing us away, tho'a good many of the younger people were amused. What any body could have found fault with, I don't see, but mortified pride can always find vexation.

After two hours, we left, the four chiefs accompanying Perry to the door; and then hastening back with joyful step as tho' relieved. Some saw signs of secret observers peeping thro' pin-holes in a side room, and I guess there were many such. On the way back, the accompanying crowd was large, & all of Napa came down, except the women, to see the show. We reached the ships at $2^3/_4$ o'clock

P.M.

June 7th

Busy all day making out Perry's note to the Regent, expressing his satisfaction at the reception, dislike of the spies tagging us everywhere, wishing him to appoint man to take the money for the supplies, & telling him his intention of going to Japan. He also got up a present for the Queen-Dowager & the other Treasurers; the former's of looking-glasses, soap, perfume &c. In the evening, took a walk up to Shui with Mayne & Dr. Smith; I was a little sore from my ride on the naked saddle I found on the horse given me at Shui, but this walk made me limber again, and we enjoyed the walk much, finding new beauties in the scenery. The crops looked well, and the whole country gave promise of sufficient food for its inhabitants. The road was occupied with many persons going to & fro, some of whom were driving horses laden with bundles. Altogether the women are the most degraded part of the population, & seem pressed down by their hard, servile work; no smiles, no laughing do we hear from them, and some of them are harridans beyond comparison. They do not flee so much as they did, but no approaches are made, apparently, to their good will. On returning, we saw some persons turning up vegetable beds with short-handled [mattoes], at a great expense of labor. A large funeral procession was leaving Shui by another road from us, and we could hear, half a mile off, the wailing of the mourners, as they dragged along, between two supporters. The coffin was carried in a high-roofed red box on men's shoulders, about the middle of the line; there were more than a hundred people in it.

The authorities made their last struggle this evening

not to take payment for the provisions furnished the ships — a strange contest, and one would wish no stronger proof of the force of law and power of espionage & oppression. However, they at last assented. One objection, that Purse Barry was not of a high rank enough to treat with them on such a matter, rather excited him, besides causing the others some amusement. It was a well arranged meeting to compel them to give way on the point, in which they have always succeeded, and which is really one of the most singular in their policy — that of refusing payment for supplies. A lot of 200 boards was also needed, and at last was promised on their part. In all these proceedings, Ichirazishi acts a most important & conspicuous part, and shows a deal of cleverness.

June 8th

A deputation was sent ashore this morning to the Mayor of Napa composed of Lieut. Contee, Mr. Barry & myself, taking with us the document prepared yesterday for the Regent, & the presents for the Queen Dowager & the two treasurers, called Man Fungming 毛鳳鳴 , the other 翁德裕 Ung Teh-yú, who manage the revenue of the other departments of the island; the last each received a sword, 4 pieces of cottons, 2 bottles whisky, 1 of wine, an engraving, & a cake. We were also to give a threatening message respecting payment if they still refused to settle accounts. On landing, I was greatly relieved, therefore, to see Mr. Spieden with money on the table at Dr. Bettleheim's house, with the purveyor making out his accounts, and all in process of amicable arrangement. Our men brought up the cash ($150 worth) in bundles of $5 or 6500 each, and natives soon carried it off. We had an easy message at the Mayor's. Mr. Contee had been at his office before, & he received us out of the door, invited us in, was much interested in the presents, so far as they could be seen; and altogether the meeting was one of the pleasantest we have had. Nothing was said of payments, but they were told that we intended to bring some cattle & sheep ashore and pasture them in the inclosure near Bettelheim's house, and wanted the 200 boards to make a fence. Many excuses were offered respecting the boards, — that they were difficult to get, as most of them came from Japan or Tuchara, and only then as dunnage, or to fill up the rice junks. I told them I have seen too many houses boarded inside as well as out, and too many pit-saws going, to think they had few boards. He then asked who was to look after the cattle, & who was to be responsible for their lives, on which points we eased his anxiety; but he made no objection to their being brought ashore that place. Inquiries were made as to where the two ships were going, and why; we also wished to know the manner of their cultivation of tobacco, & were promised some seed. After remaining more than an hour in pleasantest chat, we wished health to the Regent and all high functionaries on behalf of the Commodore, and took our departure, much better pleased than if we had been obliged to threaten them. All accounts having been settled, the pursers all returned aboard; and we may hope the authorities will make no more opposition. In fact, it is not easy to explain the reason for [~~the~~] refusing payment. I suppose that, as they themselves exact the supplies, they lose nothing by their gifts, but the people bear it all; while they deem themselves in the safest position with respect to their real rulers by adhering to the letter of the law, and considering all ships as their guests. I look upon Lewchew as a dependency of Satzuma, rather than subject to Japan, by whose prince it was conquered entirely in the

15th century, 1609. (See *Sankokf*, 177), that principality monopolizes the trade and manages the relations & policy of the island, allowing the voyage of homage to Fuhchau every year, to keep up a profitable trade and a shadow of independence among the natives. The power is wielded by the gentry, whom long usage has formed into a caste, and they sway the timid, defenceless people by a system of espionage which spreads distrust and fear of others over the whole community. The gentry maintain the spies, and are the depositories of all learning, education, and office, doing nothing to elevate or improve their serfs. Apparently, their sway is very mild, for no swords in the hands of soldiers nor even whips in the hands of guards, are seen in the streets, but it is because all resistance has ceased, and a motion of a fan or a wink is as effectual as a blow. Fear of an informer doubtless carried obedience to needless lengths, such as running away from the markets when a foreigner appears, but perhaps most of the market-people being women more satisfactorily accounts for this, and they do not now run as they did at first. There is nothing which so destroys the self-respect of the human soul as a system of surveilliance and responsibility, — constantly on the lookout that another's conduct does not involve one's self, constantly feeling that one's actions are all spied out & may be reported for punishment, you are hampered and meshed like a fish in a net, and fear to move. If the people even knew their rights, they have no power to assert them, and the only hope lies in teaching all classes the baneful effects of so unnatural a system. Whether the authorities are likely to be punished in any way for their finally coming to our demands or not, they certainly must see that we have no present intention of interfering in their internal affairs; but it is likely that a change in their foreign policy will materially influence their internal system, seeing how the two are blended; but the obvious advantages of changing the relation of host & guest for that of seller & buyer when a squadron of 1500 men come, must be apparent, even to the lowest coolie in port. Many signs of a change are already apparent.

In the evening, I went aboard one of the Japanese junks, where we were rather endured than received; there were 22 men, & they had been 15 days from Kagosima. They gave us no tea or pipes, and refused to sell Mr. Bittingar a box he was earnest to buy for a knife. The rudder post was hauled up, & lay horizontally in the cabin; and it was about 3 ft. diameter. The room was kept clean, and most of the cargo was landed. Some of the Lewchewan stchibang [sic] followed us aboard, — imps of oppression who may some day get roughly handled for their impertinence. On returning to our boat, the captain handed back a handkerchief I had previously given him.

On reaching the steamer, I found that the Regent had made his return presents of paper, cloth tobacco, saki, fans, pipes, &c., a trumpery assortment, with only a few pieces of lackered ware.

June 11th On passage to the Bonins

On Thursday morning, we got under weigh with the Saratoga in tow, & moved out of the harbor in fine style, leaving the Mississippi & Supply in port. Several persons were left ashore, among whom were Mr. Brown & Mr. Draper, the daguerrian & telegraph artists[6]; they took up their lodging in the house at Tumai. The house on the hill-top near Dr. Bettelheim's was also occupied by sending some cattle & sheep on shore thereabouts, to pasture and be taken care of, as the Mayor was informed. There is not much to do now, with the Lewchewans in an official manner, but everything in showing them the equitable & firmly just conduct proper in our dealings with them, and leading them to see that it is for their interest and peace to treat us with courtesy. Thus far things have gone on as favorably as I expected, and when the native authorities come to see that we mean what we say, they will, I hope, refrain from their own subterfuges, and treat us fairly.

Today, the poor old teacher was committed to the sea. He did not recruit at all after reaching Napa in the Saratoga, and tho' every care was taken of him on board the flag-ship, a good room & nourishing food provided, he did not recover his spirits or appetitie. He had brought all the apparatus with him for smoking opium, tho' he constantly asserted that he had none of the drug with him; I would not let him smoke, but he took it in some cinnabar-colored pills, which he called 保生丸 or nourishing-life pills, & took in large doses. He gradually failed in mind & body, and the last thing he did for me was to mark the two pictures sent to the two treasurers on Wednesday; and after that he had hardly mind enough to answer a question. He presented a sad spectacle of ghastly emaciation, mumbling and talking & moaning, now about home, and now about money. I told him a week ago, that I did not think he would ever recover, and tried to direct his attention to the Saviour, of whose salvation he was not ignorant; but he paid little heed to it, & spoke of it himself none at all. I fear his heart was never touched with a sense of his sinfulness. He died last night about 11 o'clock of inanition and exhaustion of the nervous system, delirious for 24 hours previous. He was bound up in his mat just as he lay in bed, and then sewed up in canvas. A jar of opium prepared for smoking, and all the pills he had, with a quantity of cakes, sweet meats, &c., were thrown overboard,

& his opium pipe was buried with him; he must have spent $15 to $18 for opium and other things injurious to him, and I hardly had two days' service out of him the whole time. I never saw an opium-smoker die before, and had no idea that the use of this drug so enfeebled the nervous system, and rendered the powers of mind so weak, and the whole man so foolish. He was a shocking sight, a melancholy ruin.

June 14th Port Lloyd

After a passage of five days over the most sunny seas and with the pleasantest accompaniments of breeze, temperature and progress, we anchored here this morning. The land looks native, and as if the soil was tolerably productive, for the vegetation covers the hill tops, some of which are fully 1,500 feet high. A Hawaiian, and a youth born on the island, came off to pilot us in if needed, and about 9 o'clock we anchored, almost landlocked, and deep water in some place, near the rocks. During the day parties were made up for exploring the island tomorrow, but I declined to join them in the ascent of these steep hills, lest I should not keep up. In the evening, we rambled along the beach, and visited three houses, which presented a good degree of comfort in their internal arrangements; one of them was occupied by a Portuguese, who has lived here 21 years, and has had ten children, only one (our pilot erewhile) of whom now lives with him. A daughter of his was forcibly carried off two years ago by some pirates from Hongkong on their way to California.

Wednesday,June 15th Port Lloyd

Two parties, under Mr. Taylor[Bardnard Taylor][7] & Dr.

Fahs, left early this morning to explore. I went ashore about 9 o'clock, and with Mr. Patterson, went up some of the low hills near the dwellings. All these hills had been burnt over not long ago, perhaps to cover the soil with a manure of ashes; a growth of [Carex] & [Scripus] now covered them, mixed with shrubs, all growing in the richest soil. The rock is everywhere of trap formation, containing veins of greenstone running thro' it, and nodules of iron-stone, the outer surface of which last is often blistered, as if it had been simmered before a fire; the presence of sulphur has caused this rock to decompose rapidly, and this has assisted greatly to produce the rich soil. Many parts of the soft ground were riddled with crab's holes, some of them large enough for weasels'.

The vegetation is decidedly tropical, which is rather unexpected in a place the latitude of Wanchau fú, and only 1,200 miles east of Ningpo. Here two species of palm, one of them producing a kind of cocoa-nut, the tree-fern, the plantain, papaya, sugar-cane, & pandanus, all show the tropical affinities of the flora. I found two beautiful species of Hibiscus, a Sida, of which the berry is good eating, a fern or two, and a kind of juniper. Most of the plants are new to me[8], but the variety is small. Few gynandrous or syngenesious plants came under my eye. In the damp or winter months, there is probably more variety of flowering plants in the underbrush than at this season. Few mosses or ferns appeared, the ground being grassy and dry. Seaweed is not plenty, and the species resemble moss, covering the stones at high water.

The crabs are most abundant, running over the ground, and covering the pools in the ravines, by the hundreds. They form a distinctive feature of the island, especially in the woody parts; some of them are $2\frac{1}{2}$ inches square on the carapace; along the shore, the hermit-crab is paramount, only a few others running about the rocks. In the sands, a kind of Portumnus (?) digs holes, and at low tide one can hear them snapping their mandibles with a curious clicking sound.

Few insects are seen; a butterfly, a grasshopper, ants, and sand-flies or something of the sort, comprise my list. These last are found in the dry decomposed ground in the woods, & are exceedingly agile. One lizard run across my path; brown, spotted, 4 inches long. A species of Periopthalmus was caught skipping over the rocks. The dorsal extends the whole back, the false pectorals apparently disjoined, but proceeding from the same bone; skin dark brown, black spots, eyes projecting and approaching; belly light brown. While walking over the sand, which was marked into ripples by the surf, so hard as to resist my weight; I was led to infer that the solidification of these ripples into rock, so that the layers can be easily separated into thin pieces showing plainly the original ripplings, is not so very surprising; for at this time these marks were even more solid than the shells lying on them. Probably a succession of these ripples, one above another, could even now be detected a few feet below the surface, if a large section could be removed and partially indurated enou[gh] to show the stratificaiton. The deposits on this soft sand are very slowly made, the silt coming from the comminuted cliffs brought down by the rains.

The shells are not numerous, but a large variety is produced, in & near the coral reefs, for the surf has brought up many species; the nerita, voluta, chiton, ostrea, patella and murex, have their representatives growing at low water, attached to the rocks. The coral appears very beautiful as one slowly floats over it, and the variety is considerable;

echinei are common, and hundreds of biche-de-mer, black & round, a foot to 8 in. long, lie scattered over the bottom; this species is not eaten by the Chinese. Some ray, called stingarce, force themselves over the coral; two were caught in the net, of a plain brown, with a single spine in his whip-like tail, measuring nearly a $1\frac{1}{3}$ ft. square; their mode of swimming is by an undulating flopping & rapid movement of the tail. It is a mystery to me how the spine is used for attack or defense.

There are now 39 persons on the island. Mr. Savary, an American from Mass., has lived here 23 years; two others for 21 years. Marguesans & Hawaiians are here, most of the females being of the latter. The inhabitants live peaceably with each other, but no one exercises any authority, and at times they are much annoyed by sailors. Each one shares seed with others, so that they all have much the same variety of vegetables. Turtle furnishes their chief meat, and this they salt down to exchange for provisions out of whalers. Indian corn, muskmelons, water-melons, sweet & Irish potatoes, taro, beans, onions, and bananas, are among the vegetables[;] goats, hogs, poultry, ducks & geese are reared.

June 16th

The Commodore and a large party went off to Buckland I. on a fishing & discovery picnic, taking with them the cattle & sheep brought from Shanghai, which were intended to be left here for increase. The cattle were put ashore at Williams' bay on the N.W. side of Peel Ⅰ., where they will find food, & not overrun the plantations of the inhabitants on this side. The sheep & goats were landed on Stapleton Ⅰ. which is already covered with goats, the progeny of some left there by

Beechy or some other voyager. The hogs have possession of Buckland I., here usually called Hog I. I was invited to go with this party, with the stipulation of remaining out all night, but most of them came back at evening.

In the afternoon, I went to see a cave at the entrance of the harbor, formed by the dropping down of the friable trap rock; no coral was seen hereabouts growing out of the sunlight, nor many mollusks clinging to the water-edge rocks. I suspect the direct & constant rays of the sun are necessary to these marine products. The opening is supported in front by a mass of rock, around which the water flows; it is perhaps 150 ft. high to the peak, & the water slowly percolates through, causing patches of rock to fall off. A shock of an earthquake would loosen large masses. Passing along in the boat, the coral appeared exceedingly beautiful thro' the limpid water; patches of brain, branching and a little flat, coral appeared to succeed each other; specimens of blue among the white made both look prettier, & where the branching sort covered the bottom, the resemblance to a tiny forest was remarkable. Hundreds of red [echenei], with long rays, dark purble, 5 toothed, 3 inch diameter, were seen in some places, & then disappeared, attracted probably by the food. The bich-de-mer always lay on the sand, the sea-eggs or echinus on the coral. In one cavity, a diodon was seen crawling over the bottom, and soon conveyed to our boat; the mode of inflating his body to cause the spines to project seems to be by sucking in a large quantity of water, for this one gradually shriveled as he ejected the water; yet I am told that the fish can be irritated to swell up when recently [~~cut~~] caught, in which case, the body can also be inflated with air. It is a repulsive fish, and seems uncommon in this place. Its garniture of spines renders it, as in the case of the porcupine, pretty safe against its enemies, but a shark will eat almost anything when hungry. This specimen was 10 in. long, dark brown patches on the back over a speckled yellowish-gray ground; the belly whitish.

The party returned from their trip to the other islands, giving the same report of steep hills and a few level places near the seaside. A tree was found, which the carpenter thought was mahogany. A palm, having an edible top, tasting like the cabbage-palm, was common on one side; in fact, I should not wonder if there were several species of palm here, & that cocoa-nuts would grow if brought, [~~here~~] & planted along the beach. Some enterprising Chinese would soon collect a cargo of fan leaves, if left here a month with a party, the fan-leaf palm being plenty; it is used as thatch.

June 17th Port Lloyd

Mr. Savary[Nathaniel Savary:1794~1874], the oldest resident here, is from Bradford in Mass. & was one of five men, who with a number of Hawaiian men & women, were sent[in 1830] to colonize this island by Mr. Charlton, then the British consul at Oahu; Mr. Chapin of Boston & Mr. Millichamp, an Englishman, also were in the party; the former is dead, & the latter now lives in Guam, so that Mr. Savary is, in some sort, the proprietor. No authority is exerted by him or any other person, however, and the residents live on the best terms with each other, cultivating friendly relations with each other, and acknowledging certain understood rules in respect to the capture of turtle or fish, and cultivation of ground. Mr. Mottley, Webb & Collins are Englishmen living here, & John Bravo, a Portuguese; the last-named has had ten children, & appears an enterprising man in managing his farm. The colonizing of this island thro' Charlton's agency shows that the English were early alive to the importance of the position; & he may have started the enterprise at Capt. Beechy's suggestion, after the visit of the latter in 1827. I believe Com. Perry has exercised some rights of sovereignty since his arrival, appointing Savary navy agent, taking up land and making it out, and doing what seemed to him good. If the English would govern the island, & let the coal dépôt be managed by the steam company without taxation, the suremacy & interests of the two parties would be amicably managed. The position is certainly [~~an~~] eligible [~~one~~] for a stopping-place in crossing between the Islands & Shanghai, far better than any islet we yet know of along the Japanese coast. It could be made to furnish a large supply of vegetables, and labor could be brought from China for building wharves, &c.

A record is kept of all arrivals & departures at the port, and a journal of notable events, by Mr. Savary. The number of whalers which have & are expected to visit the place this season is greater than in any year previous; two have appeared in the offing while we have been here, one of whom sent in a boat for supplies today. The establishment of a coal dépôt here would damage it as a port of supplies to whalers, whose captains are afraid of losing their men at large ports. However, they could go down SW. to Bailey I., where five persons moved from this place some 20 months ago, & began a settlement. Comparing the society now & the records of former navigators, there is an improvement in some respects; the misdeeds of runaway sailors are very vexatious, and probably cause all the troubles; one of the Saratoga's men deserted yesterday, & has not been recovered, a gain of over $200 to Uncle Sam. Ten or twelve of these characters left a few weeks ago, much to the relief of the settlement.

Near Mr. Savary's house[9], the water is shallow, & the bottom covered with fine sand, washed down from several small rills in this quarter, and allowed to spread itself over this secluded bight without being drifted off by the tide; the drift in some of these rills shows that at times violent torrents roll along the beds. Here the net brought up large quantities of mullet and a species of silver perch, which finds their food in this silt. Pieces of dead coral are scattered over this patch, doubtless mixed too with remains of fish, land crabs, tropical and temperate vegetables, and trap rock, all washed into the basin. If this piece should be raised up as sandstone by some volcanic force, what a miscellaneous assortment of relics would be found imbedded in it as petrifactions, such as would puzzle the most acute. Yet we can see how natural is the process by which such heterogeneous constituents have been brought together in this spot, while in other parts of the harbor there is nothing of the same kind, coral covering the bottom in all but the deepest holes. It is by observing how deposits are made now, that the mind can come to a clearer idea of the manner in which older rocks have been formed, and their constituents brought together. The various things found in the trap rocks of this island seem to have been deposited in an irregular manner, for in one place we have viens of iron-stone, in another of quartz in different forms of chalcedony, obsidian, greenstone, in a third a bed of iron pyrites, then sulphur without the iron & discoloring the rock which it causes to decompose, and again a fine grained isomorphic, half-crystallized basalt, approaching the columnar form. Almost all the mountain rivulets are sulphureous, some of them unpleasantly so. Is the tropical type of the flora owing to the heat of these rocks being increased by their proximity to the internal fires which crop out at Volcano I. north of them? Earthquakes are not infrequent, [~~here~~], but no data concerning them were obtained; they have not been very severe since the residents now here came.

The scenery of this group is imposing, the peaks rising sheer up in the steep [~~peaks~~] points which show their origin. One of the exploring parties suddenly found themselves on the brink of a cliff, fully 500 fet. down. Most of them are susceptible of a growth of grass and vines, but not one acre in a hundred can be cultivated. From the deck of the Saratogra, one summit behind John Bravo's house bore so strong a resemblance to a lion's head & shoulders, that we said "John Bull must have the claim to prior possession as his seal was on the mountain".

Many species of shells might be collected in a short time, if one would search & drag for them. Species of Arca, Chama of large size, Cyprea, Conus, Patella, Nerita, Chiton, Anomia, &c., are frequent; few oysters, & not many land or lacustrine species. Fish are plentiful, but the inhabitants find turtle to be more profitable game; species of Diodon, Balistes, Serranus, [~~Lophius~~] Tetrodon, Shark, ray, Mullus, and Perca, have been seen, some of them abundantly. Crawfish, some of them $3\frac{1}{2}$ ft. long, are common; two species were brought us.

One of the pleasantest sails I have had was taken this evening after sunset; Mr. Madigan & I took a canoe, and paddled to Mr. Savary's, where we remained an hour. The row over the smooth water, in a bright moonlight, which made a beautiful contrast of shade and moonshine along the banks and thro' the harbor, was pleasing to me, only recently from the hot bricks of Canton, & I enjoyed it greatly. All these canoes are hollowed from single trees, with a bulwark added to the wale, & furnished with outriggers & sails; for the uses of the islanders they are better than a boat, & are easily managed by one person.

(For further notices of the Bonins, see Klapzoth's translation of the San Kofk Tuzan, Beechey's voyage of Blossom, Ruschenberger's voyage in Peacock, Chinese Repository, Vol. Ⅲ , Siebold's maps & memoirs, Lütke's voyage.)

June 18th

Taking the "Saratoga" in tow, the "Susquehanna" steamed her way out of the harbor this morning, the same fair weather attending us which we have had for the last decade. The Bonins were soon lost to sight, and no very dear memories left behind, if the complaints of the officers respecting bad washing at high prices & few provisions at extravagant rates, could be deemed an index.

However, the people did their best at washing, & sold us what they had, doubtless taking advantage of the rare chance of a ship of war to make the most; but they would be blamed anyhow, let them do what they might.

In the afternoon, the island of Rozario or Disappointment I. was passed; a low coral island, probably, once two islets, & now joined by a shingle beach of coral fragments. The serf beat up fully 30 ft. high as we passed; the highest point of the island was hardly 50 ft. high. Reefs defended it wherever we could see it.

June 19th Sabbath

Owing to squalls, there was no service today; and the general appearance of the ship's company differed very little from that of other days. God is not reverenced in his holy day on board ship, to a great degree. What misery it would be to most of those here to be shut up in heaven, forced to join in the anthems & praises of the blessed, and invited to spend eternity in keeping God's holy law, for which they have here no relish.

Wednesday, June 22nd [This page is partially tattered,which makes it hard to read.]

Our pleasant SW. monsoon weather still continues, and we get along 6 to 7 knots an hour over smooth seas, having occasionally a favorable slant of wind, so that the sails can be set. At noon, we passed within miles of Borodino I., on the North of us, a low, coral island, not over a hundred feet above seawater, & covered with vegetables & trees; it consists of two islets, the largest 6 or [7] 5 miles long; the smallest a mi[le] lying N.E. of it. The surf broke over the reefs along the whole length of it, and there is probably no very safe anchorage near the shore; and so far as could be ascertained no inhabitants, either, but no conclusions could be safely drawn from such a view. It lies in such a direction from Lewchew, that it is not unlikely that the inhabitants have been to it, and may still cultivate it. A good survey of the two would be well worthy of being made, not only to ascertain its capabilities for sustaining a population, as to s[ee] if there was any shelter there for a vessel in distress. It is the only land between Lewchew & Bonin's, on which any person could find retreat, or resort to in case of shipwreck with any hope of sustaining life.

Thursday, June 23rd

We anchored at Napa about 5 o'clock. P.M. and found t[hat] the Plymouth only had arrived; she brought news from Shanghai & U.S., & I was gladdened by hearing of the health of my dear ones at Canton, which I had not much expected. Also, of the provisional escape from drowning of Dr. Parker [Peter Parker:a medical missionary,ABCFM;and in 1855 appointed to U.S. Commissioner to China][10] in the wreck of the steamer "Larriston" on Turnabout Ⅰ. near Hactau, in his passage back from Canton, for all which I desire to be thankful. Many were disappointed in not getting letters.

Dr. Bettelheim's presence was soon announced on board, but he had not much to communicate. He thinks the northern part of the island ought to be searched for coal; I think there would be as much chance for finding gold as coal in this islet, & who is to dig it? After he had gone, two officials from the Mayor of Napa came to hand in his card to the Commodore. They were desirous to ascertain where we had been, but their knowledge of the world around them is too limited to know even where the Bonins are situated.

I have been reading an abstract of Levyssohn's recent publication on Japan[11], in which he endeavors to excuse Dutch servility and Japanese seclusion, showing by the way that there is very little prospect of a successful termination to this American attempt to open trade & intercourse with the islanders. However an attempt must be made some day or other, and until the temper of the government & people is ascertained in view of a stern demand from abroad, how is any course of action to be marked out. The opperhoofd's [VOC opperhoofd, Dutch=supreme head= 商馆长] views are as contracted as the little island of Desima, where he has resided; no reference to the general interests of humanity, to the pitiable heathenism of the Japanese and their ignorance of the revealed will & laws of their Maker, to extension of intercourse & consequent elevation of character, or to the diffusion of true Christianity among them, is to be found in his pages. It is, to my mind, a fair example of the influence of sordid trade on the human heart.

June 24th

I was engaged during the morning in making out cards to send to the leading officers of the government to dine on the flag-ship on Tuesday next, the prince, the Regent, three treasurers & mayor of Napa, telling them in oriental style that we "had prepared goblets & awaited the light of their presence" at 4 P.M. I took them to the Mayor's office, where I learned that it would be necessary to change the Regent's card, the old one having been made to resign or been deposed, while we were absent. One is inclined to speculate as to our agency in the degradation of this imbecile man; the last paper he brought aboard ship on the 4th, much to the surprise of all, intimated to Perry that he (P.) had his fate in his hand, that he could not allow us to go to the palace, alledging however only the illness of the Queen Dowager, and the commands he had received from the 國主 "sovereign of the land" to entertain the Am. commodore at his own official residence; then, his very undignified act of remaining out in the street of Shui to coax or invite us into his house, may have been a last effort to avert his probable fate, & show that he had done all he could to prevent the entrance to the palace. However, no charge could introduce a greater non-entity than this man seems to be, for he is the most of a child of any officer we have had intercourse with. Nor can one feel much sympathy for men who put themselves at the beginning in an attitude of mistrust, reserve, and distance, refusing that intercourse which unfettered humanity would take, and deriving no benefit themselves from this churlishness; such rulers as these curse themselves and their people.

Be this as it may, some causes have overthrown the Regent, and a new man may be free to take a new course. The card to the Prince was at first declined on the ground that he could not come, but I would not hear to the excuse; he is said to be 12 years old, but why they style him 太子 or heir apparent, I do not know, if it be true that the father is dead.

In this office of the mayor's is a tablet showing the influence of Confucius' maxims: 孝悌尧舜之道不外乎此 Filial duty & brotherly love: the doctrine of Yáu & Shun are nothing but these two. The mayor was desirous of ascertaining when we were going away, & where Perry had been; to the first I pleaded ignorance, & endeavored to answer the second as well as I could, which was not easy without a map. Perhaps my answers would hasten the dispatch of the junk lying off the Roads, & this may explain their earnestness. Ichirazichi shows great tact in the way he manages his questions, and I suspect his influence is proportional to his parts.

In the evening, I went around to the house in Tumai, and found that it had been made a much more comfortable house than the other could ever have been made, for it is, larger, has a better yard, & is cooler. The other is now actually occupied as a schoolroom, as we ascertained by going into it, where we found 28 lads conning over a Japanese edition of the works of Mencius, just as if they had been in China, squatting about on their haunches or jumping

[~~about~~] around the room. Even with all their childish glee, there was the same serious air which seems innate to a Lewchewan; Mr. Spieden says he has only once seen the people laugh heartily, & that was when they felt the shock of a galvanic battery on board the Mississippi.

Rambling over the hills back of Tumai reminded me of the walk which we took in 1837, (having Mrs. King & Capt. Ingersall in company,) in these parts, tho' I am not able to recall the locality at all. We went up to a Budhist temple to see what could be in the building, and found a party of priests sipping tea & smoking; a sacrifice of cooked dished was spread over the main room, arranged on low tables, in front of the idols, having lamps burning. The party gave us a cold reception, motioning us out of the house, and refusing us an entrance into the temple; indeed we could hardly get a drink of water, and did not tarry long. The Hib-rosa-sinensis[12] was in full flower in the yard, which was kept neatly. Almost all these temples, I am told, have an adjoining building, for the entertainment of guests & travelers, who are thus enabled to house themselves without incommoding the priests. The location of this establishment is very pleasant, and everything around it was riant and peaceful. May God in his mercy soon change the sullen superstition of the inmates to a joyful faith in his Son.

June 25th

Most of the day was spent on board the "Supply",where I went in the morning to go with Dr. Wilson and examine coral beds, but found the tide so high that we had to wait till afternoon.

On reaching the coral reefs, we had some difficulty in keeping the boat easy, but by the men getting overboard, may pretty specimens were obtained of madrepores and other sorts, with two kinds of echinus. Hundreds of the blue coral fish were flying from one hole to another, their bright skins alternately showing blue & green as the light was reflected from them. We came across one agile fish, which seemed to walk along the bottom, and was perhaps a siren. We carried the coral ashore to bleach it in the sun near Dr. Bettleheim's house; at his house, a large number of natives were assembled, looking with some interest at a pile of condemned biscuit sent ashore, afraid to touch or take it in presence of their overseers.

The people havc a pretty mode of planting trailing plants run along the top of the walls around their houses, both to mat them firmly & raise a defence against climbers. Bastard banian, caeti, bamboos, organines (Murraya), and a sort of ivy, have all been seen.

In a funeral procession which passed near us, the bearers of inscriptions to propitiate the gods of the way took the lead, then a company of well-dressed men, all clad in brownish white dresses, & then the male mourners blubbering & crying as they stumbled along, half borne up by assistants. The coffin was inclosed in a bier, formed of a tray and a cover which completely concealed [~~the~~] [~~coffin~~] it; the whole was red, and was borne by four men, who showed that their burden was not a light one. After the bier, came the female mourners perhaps 30 in all, some of them friends supporting the crying, wailing women, and all protected from the crowd by men carrying a net on each side stretched on poles. There was no music, & the red bier was another deviation from Chinese custom. The graves in this vicinity are substantial erections in the same general style as the Chinese tombs about Canton, resembling a letter Ω, or else an opening into the rock [~~in~~] thro' which the coffin is thrust into a recess, & then closed with masonry. Considerable labor has been laid out in scarping the ledges in many cases to make a face for the tomb, or in building a wall to inclose a small area in its front. No inscriptions have been seen on any tombs, in which they differ from Chinese; & I suspect their sepulchral rites partake more of Japanese customs than Chinese.

June 26th

Altho' it was the Sabbath, Ichirazichi came off to the ship to intimate the acceptance of the invitation to dine with the Commodore on Tuesday. He made inquiries where we had been while absent, & I got an india rubber globe to show him the position of the Bonins with respect to U.S., China & his own country, and strongly impressed it on him that his gov't must expect to have many visitors coming into their ports, and the sooner they were treated properly, and supplied with what they needed, the better it would be for his country. He wished to know why boats had gone up to Port Melville, as letters had come down from Unting stating the arrival of boats there last evening; I told him they were sent to survey the harbor, & would return in two days; & that we intended to go everywhere on their coasts examining the shores, so that ships might know where to anchor. In respect to everything relating to foreign intercourse, and the courtesy due to ships, I give these officials no comfort or hope of a better time coming; they are now learning their duty in the gentlest manner, & must understand that we are in earnest. The report that Sháng Tá-mú has ripped himself up is gaining ground, & excites no little displeasure among some as one of the sad results of our course; but I have great doubts about it, and if it were so, the execrable laws which compel such a step are more to blame, in my view, than we are, who had no idea of such a contingency.

Dr. Betteleheim came aboard after his service was over in the Plymouth, and made himself somewhat dubious by the way in which he spoke of the succession to the Regency, & the fate of the old one. This same Dr. B. contrives to heap a deal of ill-will & contempt up against him, by his conduct.

Monday, June 27th

In the evening, rambled over the reefs with Mr. Jones, collecting fish & mollusks, all of which were drowned in my jar by mistake. In the night, the crew of the Mississippi gave a theatrical performance to the squadron. The commodore rather favors these things, saying that their effect is to keep the crews in good spirits; the men are pleased enough to have time given them to learn their parts & paint the scenery, a sort of shirking their work which others do not like.

June 28th

The arrival of the "Caprice" this morning gave unwonted stir to our little fleet, and the letters, parcels, stores, &c., were soon scattered among their respective owners, a Chinese assistant to take the place of my old man Sieh, & a servant-boy Alai to attend on me, falling to my share. They both talk the dialect of Shánghái, and I am likely to become expert in the court dialect before I get home, as this teacher needs a deal of explanation. I was glad to see Capt. Maury look so well, & think he has given satisfaction.

Towards noon, the Commodore began to fidget concerning the arrival of his guest, and as the rain came down briskly, it was in a measure doubtful; the boats were sent according to promise for them, but Betteleheim's fears added to the uncertain state of the weather, induced him to send us both off also; we met them all aboard the two cutters, and had our row in the rain for nothing. Betteleheim was cross too, because the Regent was ahead of him, & hallooed to the boats in vain, making me wish I was out of company.

The new Regent, Sháng Hiung-hiun 尚宏勲 two of the Treasurers, and the mayor of Napa, with many té-fú, or subordinates, in all 18 or so, came off. Capt. Buchanan took some of them over the ship and into the Engine room, and I went with others elsewhere, but there was no time to show them much, as Perry hurried all down to the table. He seated the Regent & a Treasurer on his right & left, the other two were at the opposite. The Regent has a family likeness to the former, & acted in the same still, hushed manner, exhibiting more uneasiness and constantly glancing here & there as if afraid of treachery. The others enjoyed their dinner & wine, tasting of all & clearing their plates often. The Regent thanked the Com^re. for the cattle, and promised to rear them; he was further promised some seeds from U.S. to distribute among the people. He had brought some saki & sweetmeats off himself, which were laid on the table too. While dining, many sorts of spirits were drunk, and Betteleheim evidently acted as if under their influence, getting up & sitting down, talking & gesticulating in a strange way. I wish more pains had been taken to inform these officers than to guzzle them, but darkness was coming on, & no time for aught but eating. The Regent rose to leave two or three times, but was motioned down as often, his host perhaps forgetting that at Shui he left long before the last course, and had not the same excuse of night coming on. The Regent was told that we were going to Japan soon, and that other ships were coming here, & we hoped friendly intercourse would spring up. The health of the guests & their country was drunk, in which they joined, but proposed nothing themselves; indeed, nothing could interest or please the Regent, except to get off. The rain came down so fast, that after the guests were on deck, they could not go, and went into Capt. B. cabin to rest awhile. The marines were marshal[l]ed & the band played, so that nothing was wanting, to show them respect; I suspect the attendants got very little to eat, tho' their eyes & ears were filled with sights & music. I tried to ascertain from the interpreter whether the old Regent was in Shui, but had no chance; Bettleheim thought he was imprisoned or banished, and increased the dislike of some to him by the smirk with which he told of the poor man's fate, a fate which I think is doubtful. I don't much wonder at his feelings, however; living here for so many years, and deprived of common comforts thro' this man's means, it is not surprising that he should wish a change of rules. The party of Lewchewans left at sunset, but he remained to try to settle accounts with the purser or caterers, and nearly got a discharge from the ship by accusing the officers of cheating him. It is strange to hear the dislike felt against him by the squadron, yet I can explain it mostly without deeming him to be a scoundrel as others do[13].

June 29th, Wednesday, Napa

Dined with Wayne & Maury in the Caprice, & then took a talk to Shui with the latter, much to his delight, as he had not been ashore before, & we really had a pleasant walk through the charming country. We went over to Wí-dumai, the embowered village, and returned along the seaside hilltops, from which the view was the one McLeod describes[14], a mixture of sea & shore, copse & wood, cultivated patches of many colors checking in the whole, and graves of solid masonry placed in grassy hill-sides or surrounded with solid stone-walls. The palace grounds at Shui indicate much taste, & the rivulet which runs by Tumai is there collected into a pool of ¼ sq. mile or so, affording many conveniences to the people. We met the tallow-tree, mulberry, lotus and taro, cultivated, but not to great extent. The people run from us, & one left a pail of cool water in the streets for our enjoyment. The strata of limestone is lost sight of as one ascends to Shui, where granite alternates with it.

When we spoke to people this evening, they would put their fingers in their ears, — a new device to hinder intercourse, which those who did it rather laughed at, for we saw a lurking smile on the faces of several at the grimace they were told to make.

June 30th

Coaling ship all day, which makes the vessel uncomfortable, in spite of all the precautions taken. The "Brenda", which came in Tuesday, is discharging her load into the "Mississippi", and every preparation making for a start. The Supply is to remain, keep possession of the house at Tumai, and the Caprice is to remain at Shanghai just long enough to be back here by Aug. 1st.

I have been busy translating the President's letter, and find my Chinese assistant a mere office copyist, one who has had but little readings, & is not quick at catching my meanings. Added to this, his pronunciation differs from mine considerably, so that we are frequently thrown off from catching the meanings. He is good-natured & patient, in which qualitites I can learn.

July 1st

Went ashore this morning to carry a lot of seeds to the Mayor's office for the Regent. I had a long talk while there, chiefly to answer a petition received from the Regent the day before through the Mayor, who came on board the flag-ship to present & urge it himself. The purport of this paper was that the Regent requested Com. Perry to send back two Chinese who had been sent over from Shanghai in the Brenda to Dr. Betteleheim as assistants, as they were not wanted. I told the officers that we had no hand in bringing them over, that Dr. B. was an Englishman & these Chinese were sent by English officers to him, & that we could do nothing in the matter, adding, that they had better give up all such ideas of preventing people coming to their shores to live if they wished to do so, and they [*sic*] sooner they began to treat foreigners like friends, allowing them to trade as they pleased, not ordering the people to run from them, or the women to hide themselves, the better they would get along with them. They seemed to understand the matter, but I suspect are not very free to follow what is advised. The personal position of the Regent when he went down into the engine-room, urged on by Capt. B., and terrified at the ponderous machinery before him, is not unlike his political position now; pressed on either side by fear of China & Japan, urged to change by what they begin to see is a power more irresistible than either, and yet not seeing their way to do so very clearly, the rulers here deserve more consideration than all have given them. I told them that henceforth, America & other ships would visit them more frequently than before, and would expect to be well treated. We had treated them kindly, and expected to get similar returns. Ichirazishi[15] was very particular in his inquiries as to what ships of the squadron were coming, which was to stay, what force was to be out next year, and other questions showing the desire to ascertain all our movements. I told him all I knew; and furthermore thanked him, on behalf of Perry, for building the tomb over the body of the boy buried from the Susquehanna, June 3rd. He said it was their law so to do, & I commended such a custom. The interview was quite long, but I hope these officials are beginning to understand that we are friendly if they are, and that we mean all we say; to me, they appear like school-boys who need some threatenings and coercion for their own good, to show them that nations have mutual claims, and they must acknowledge these claims. But what can weakness & might, such as are here in contact, do? We are our own expounders of what we wish them to consider right; but they are not able to see the matter from the same position. However, during the last six weeks a good beginning has been made in this instruction, no harm done to them, and proof enough given of our intention to take all we wish if they are slow in granting it; they have derived some benefit I hope, tho' I fear there are more lesson in this political economy still harder.

At parting, I received some pipes & fans, & some tobacco-seed, and the good wishes of the company. May they soon be made willing to receive the gospel.

The Caprice sailed at noon, sooner than I had supposed she would. Dr. Bettelheim has so tired out the officers, that few showed any warm desire to help him get this letters off, & he was too late; yet there is much to be said on his side too, troubled and vexed as he has been with provision bills from every mess in the fleet. In the evening, went to Wi-dumai for the third time; the people were more friendly than ever, & the village looked charmingly. The scenery hereabouts is truly charming from its peaceful character, evidencing so much the quiet character of the inhabitants, and one cannot fail to relish it.

Monday, July 4th

We sailed from Napa on Saturday morning, taking the "Saratoga" in tow, & followed by the "Mississipi" having the "Plymouth" in her rear. We have sighted several islands lying NE of Lewchew, some of them not accurately laid down. Today has been a holiday, and a salute was fired at noon from all the ships; this outburst of patriotism did well enough to announce to these remote waters the coming of the universal Yankee nation to disturb their apathy and long ignorance, and I hope there will nothing worse come of our visit hithe[r]wards than firing some salutes and making a noise. I pray the Governor of nations to so prepare the hearts and allay the fears of people we are visiting, that this mission to them shall be as peaceable as the [tenor] of President Fillimore's letter to the Emperor, and that their sovereign and his advisers may be led to entertain these proposals favorably. I am sure that the Japanese policy of seclusion is not according to God's plan of bringing the nations of the earth to a knowledge of his truth, and making the Gospel of Jesus universal as a guide to eternal life; and until it is broken up to a greater or less extent, those purposes of mercy are impeded, for his plan is made known and we have no knowledge of any other. To immortal glory at his right hand, the Japanese can have no entrance so long as they are idolaters; and as it is by preaching that the plan of salvation is to be made known, their policy must change to allow this promulgation. If they have known Christianity only through the medium of Romanism, and have never had the means of studying the law of God for themselves, but regarded this new-fangled doctrine as only a cover for political schemes, it is to be lamented, but seems to me to present no good reason why the nation should be left still in seclusion and ignorance. With regard to the real views of the U.S. gov't, or the plans of Com. Perry, I have less confidence, since I have seen more of his character; and the previous experience of victory in Mexico may strengthen his determination to drive matters by force which could not be attained otherwise except by long patient treating. However, let us hope for the best till we see the worst, and trust much to the overrule of God to make the wrath, the ambition, and the pride, on man to praise him by advancing his glory. How this can be done, we can see better than now, after his plans are seen developing and the power of opposing forces made manifest. There may be a war, but this does not seem probable from what I can see of the position of parties, so far as I can now judge. It is a matter of rejoicing to know that God does rule the affairs of all nations, and this rule will be exhibited here as it has been elsewhere.

July 8th

Land appeared on the N.W. at daylight, thought by some to be C. Toötomi, and erelong C. Izu was seen; a chilly air showed the promixity of the mountains, which appeared in the distance about 8 o'clock. Many junks were seen near the coast, but not many in our route. The islands lying S.E. of C. Izu toward Tatsisio I. showed less plain owing to the morning mist than when I was here in '37, nor was any symptom of volcanic action seen on one of them; the sea was & has been clear of seaweed & pumice, until this morning a little was seen of the former. We distanced whatever junks were bound up to Yedo, the two steamers going thro' the smooth water at 8 knot pace, and across the Bay of Kawatsu, between Capes Izu & Sagami, almost no boats were seen; one small craft seeing us coming up rapidly, took in sail, turned about, & pulled away for Vries I. [Idzn O-sima][16] as if its existence depended on their haste, doubtless to comfort the inhabitants with tidings of the happy luck they had had in not being run over last night. Mt. Fusi rose in the distance beyond C. Izu, with its bifurcated peak, accompanied by many other less elevated points, but all of them concealed more or less with clouds; the mist concealed the coast, and hid us too, probably, from the people. The remarkable white rocks along the coast were hidden by the same cause; but a few guns, which were ordered to be scaled, made our presence known, perhaps, to those who could not see us. The sight of land diffused a feeling of exhilaration through the whole company; and certainly the dim idea any of us could have of the results of this visit upon us or the Japanese, was calculated to excite our minds.

The ships anchored off Uraga about 4 o'clock, the two steamers being nearest the town. Many boats like scows, full of athletic, naked boatmen, came near. I asked one well dressed man in the nearest to the gangway, to send ashore and request a high officer to come off & take a letter to the emperor. While talking, a second official came up, saying,

"I talk Dutch," whereupon Mr. Portman[17] told him that the Commodore only wanted to have a high officer to come aboard; he then pointed to the highest one there was to take such a commission, the 2nd governor in Uraga, standing near him, and that he could not venture to go ashore for any other. After some parley, these two were admitted, & received by Lt. Contee in Capt. B.'s cabin, and told that the President had sent four ships on a peaceful errand to the emperor with a friendly letter, which it was desired to send up to Yedo with dispatch by a proper person. No answer was given to the questions made about our course, men, equipage, &c., which they were told national vessels never described. The town of Uraga was said to contain 1,800 houses, & it was 18 ri or 27 miles to the capital. These officials said they would come tomorrow & receive the letter. The "commandant", as he called himself, had writing-paper brought & made a report in official form of what he had heard, which he read to the interpreter, & then took leave. He was enjoined to send all boats away, as we would not go ashore, & they were therefore useless; this was done to as great degree as one could expect, as soon as they went away. Both these men were dressed in black crape upper cloaks and a sort of petticoat, having the coat of arms stamped in white on the arms & back; their long swords were taken off as they sat down. The commandant showed his official insignia, a kind of brass trapezium with a swinging vernier, the rim marked in Chinese figures; he had written [role] containing commands ordering us, as I suppose it would all ships to whom it was presented, to anchor where we were, but he did not offer to show it, as we were already anchored.

Our position was above that of the "Morrison" or "Columbus",[18] and it commanded the town; four rockets were sent up before anchoring from Kin-na zaki, the point seen above the town, probably to inform the capital. The town lies closely to the beach, many boats lying off, and appears compact & well built; four forts are near the shore in various places. Most of the boats near the ship bore small square flags marked others , both said to show they belonged to the gov't; no arms were seen in the boats, but many well dressed persons had come off to see the ships, and I was somewhat surprised to see them go ashore with so little apparent reluctance, when we told the commandant to order them away.

The bay looks [~~just~~] as it did 16 years ago, and the reef of rocks is as I remember; we did not see the town then as we can in this position, but the headland round which we saw boats come & go so often I remember well. The authorities will bring no guns now to drive us off. The coast line from Cape Sagami is well defined, a steep bluff, with little beach, well wooded, & cultivated here and there, trees along the ridge — these are the features. No preparations of a hostile nature are visible, nor do the forts appear well mounted or manned; nothing is to be seen of all that Bettelheim was so confident of.

About 6 o'clock, the two officials came back with a third, and were received as before. They made a long talk about the necessity of taking our letter to Nagasaki, the only place where Japanese laws allowed its reception, & that the governor on shore could not receive it; we asked them if he took the responsibility of refusing it, and that having rec^{d}. our orders to go to Yedo from our own ruler, we were as much obliged to obey as he was; further, that he had told us on the first visit, that he would come off tomorrow with a higher officer to receive it, & that he must have known the laws as well then as he did now, two hours after, and if he did not come & get the letter we must take it ashore ourselves. These replies rather cut short their long talk, and they agreed to come for the letter to-morrow, as they went over the side. Before leaving, the sharp faced commandant went off to look at the big gun, asked if it was a Paixhan, took its range to the shore, and then examined the locks of the guns near the gangway; he had evidently a commission to this effect, but we gave him no chance to see much, for we have an object highly desirable to effect as peaceably as possible, — that our letter be received without force, so that there be no collision before the gov't is fully aware of our designs. I pray to God to order these combustibles now brought together so that they shall warm each other rather than mutually consume one another.

July 9th

Watches were kept during the night on board as if expecting an enemy; and on shore the tinkle of a bell or gong was distinctly heard during the whole night. Several boats full of men were lying off shore at daylight, so that it is not unlikely that watch and ward were maintained by both sides while darkness reigned, and the sight of something like black screens along the shore strengthen this idea. About 7 o'clock, the highest officer at Uraga, named Yezaimon, attended by two interpreters and four or five others, came off; a parley took place off the gangway, as to the object of the visit, rank of the officer, and person they could see. At last Capt. B. was ready to receive them in his room, three only coming up. When seated, Yezaimon stated that he had come aboard to express his official inability to receive the letter, and though he himself was willing to take it, the laws of the land forbade it. It was replied that the ships would remain here till the letter was received, and that we wished to have a suitable person come aboard to take it; that we had been sent by the President to the Emperor, and must execute our commission, which weighed upon us as strictly as their laws did on them. Reference being made again to Nagasaki, they were told that we were sent here, and because it was near the palace. The originals of the letter & credence were then showed them, and also the package containing the translations; they showed little or no admiration at them, but wished to know the reason for sending four ships to carry such a box & letter to the emperor, yet whether the reason assigned, "to show respect to him," fully met their doubts as to the reason for such a force, could not be inferred from their looks. A courteous offer of water & supplies was made, which was declined, and Yezaimon added then that he would not come off again before the termination of the four days allowed to send to Yedo, a period they themselves set as to the time required to send up and deliberate upon the matter. They were clearly informed of the meaning of a white flag, and also that visits were out of season till after the flags were hoisted in the morning.

During the whole of this interview, the bearing of these Japanese was dignified and possessed. Yezaimon spoke in a clear voice, and thro' Tasunoski, who put it into Dutch for Mr. Portman, I could make out almost all they said, but it would require considerable practice to speak that style, & I am not sorry that one of them knows Dutch so much better than I do Japanese, for I think intercommunication is likely to be more satisfactory.

At the close of the interview, the interpreter said the officer p[reviously] sent was the highest in Uraga, & his name Yezaimon; "What is the name of the captain of the ship?" He was told, and nothing could be more polite than the whole manner of this incident. While I was on the gangway before they came up, one said, "Are you an American?" — "Yes, to be sure I am," I replied in a tone to intimate some surprise at the question, whereat there was a general laugh. Tatsunoski then asked my name, & I his; Yezaimon had a brocade pattern of drawers, but a beautiful black gauze jacket, a kami-shimo, I suppose they call it. His crest was on his lackered hat also; the boatman had a blue & white striped livery-coat, & looked more decently than the naked fellows yesterday. A flag with 三 marked on it was explained to denote his being of the third rank. Among his attendants was one red-cheeked, girlish looking young man of a prepossessing features. A large [buccinum] was taken out of a box, adorned with tassels, and having a brass at the vertex, but I could not make out its use. How curious one becomes when allowed to see things & people by glimpses in this way, and unable to ask & explain fully!

We are anchored in 21 fath., and off Kan no zaki, 43 were found, so that we can further up on occasion. We are fully 4 miles about Ingersoll's anchorage, & have the peak of Fusi san visible over Uraga, or Uraka as Siebold's map has it. On the opposite side of the bay, two considerable towns are seen, one of them a resort of boats; the land rises gradually in that direction to no great elevation, but seems to be rather well cultivated. No boats are about the ship, but numbers are sailing in all directions, some of which evidently pass near the ships to see them. The tide runs very strong, and various patches of seaweed & medusa are common. The bay is a fine one, & Mr. Heine has taken a drawing of the shore above & below Uraga. Four forts are hereabouts, one of them a recent undertaking, but they show few guns mounted, and no strength. Parties of soldiers are stationed on shore to watch our landing, and one boat came so near as to start them up, to defend their inviolate territory.

July 10th, [Sunday]

Little of interest occurred today. The two interpreters came alongside with a new officer, described as being of less rank than the others two, whom we have had on board before. As he had not come to see or say aught respecting the reception of the letter, but on some other business, to explain which he wished to come up, he was not allowed to cross the gangway. The boat bore two flags, one the usual white-black-white one, & another with a figure 5 in red; the men had the blue-white striped jackets we have usually seen; the order and discipline maintained in these boats is superior to Chinese boats. Many boats, bearing various flags astern have gone about the ships from time to time, evidently to gratify curiosity; perhaps high dignitaries have come from Yedo to see the big ships of which rumor probably gives full accounts. Soldiers are evidently collecting in our vicinity, and glasses are so constantly in use that no movements of importance along shore escape notice. Trade has not been suspended at all on account of our presence, for the bay is at times alive with boats, and some 60 were counted today passing up northward.

All these notices and interruptions tend to distract one's thoughts from the seriousness of the day, which except the formal service at 10½ o'clock has hardly been referred to as being different from other days. I think to lead a life of godliness on board a man-of-war must require a large measure of the Spirit.

July 11th

A surveying expedition was fitted out today to explore the bay northward, consisting of a boat from each ship, and the "Mississippi" for an escort. They started about 9 o'clock, and the boats were erelong out of sight around Kan na zaki, where the Japanese had collected many boats, each containing 8 or 10 soldiers all accoutred, & carrying lances & swords, their banners flying and officers stationed, to intercept them. Mr. Bent's boat was nearly surrounded, and if the steamer had not been at hand to support him, he would perhaps have been attacked, & doubtless compelled to return[19]. Swords were drawn, but the Japanese were content with demonstrating their purpose, and drew back as the party came on. About 45 boats came out against them, quite enough to have turned them back; no matchlocks or cannon were seen, but may have been concealed. Some officers wore brazen helmets and a sort of cuirass, and some had red jackets. [~~There~~] [~~was~~] [~~a~~] A boat came near the steamer on her return containing an officer or two in rich dresses, but no intercourse was had with them. The boats found deep water about 10 miles, and it is thought the city of Yedo was seen in the distance. Great numbers of troops were seen embarking from the low land northeast of us; and beyond the same spot a large city was seen, perhaps Imatomi.

While this party was away, Yezaimon & the interpreter Tatsunoski came off, & after being seated in the cabin, & compliments passed, he told Capt. B. that it was probable that the letter would be received tomorrow; & that if he came off, it doubtless would be taken. We expressed pleasure at hearing this, reiterated our amicable intentions in coming here, and told him we expected that his gov't would receive us in a friendly way. The real design of the visit then was hinted at by an allusion to the steamer, and they were told her object simply was to sound the bay so that if we came here again, we should know where was the proper anchorage; & that she was to return in the evening. The two gentlemen were in good spirits, took a glass of wine, and seemed pleased at the offer of examining the vessel when they come tomorrow. They soon rose to leave, and were unusually polite at departure; one of their flags had a figure six on it. Some of the flags seen ashore & red-jackets, too, today, had 會 on them.

July 12th

The appearance of the bay this morning was beautiful from the sun shining through the mist which lay thinly on the water, and thro' which the shores were faintly visible; the whole was carried off by the rising sun. Few vessels were stirring before nine o'clock.

About 10 o'clock Yazaimon, (whose full name is Kayamarin Yezaimkon 香山(カヤマ)連(リン)榮(ヤ)清(ザイ)門(モン), with an addition of 永考 (Naga-nori,) & the two interpreters, came in a large boat, to say that the letter would be received, but that he could not tell exactly the day. This led to explanation, and I was not surprised to see that in their minds the copies had been confounded with the originals, and that they referred to the latter and we to the former; that they had made an appointment of an envoy to take these, while we supposed them to be hesitating about the transmission of these. The copies were shown them, and Yezaimon refused to take them, preferring to make further application to his superiors, to learn their will. The conference was very long from the apprehensions of our visitors, and their constant reference to law, so that at last the Commodore sent in his note, that he would never go to Nagasaki nor receive aught thro' the Dutch or Chinese, that he would deliver the originals only to an officer of equal rank or to the emperor, and that he must see his credentials. It was assured us that the envoy was a high officer, & I suggested that he was the prince of Sagami, in whose jurisdiction, Uraga lies. A proper place was now preparing for receiving the letter, for there was no public hall suitable in such a place. The need of first receiving the copies was insisted on, and that it was indispensable to meet an equal; so, after three hours' talk, and receiving a paper in Dutch with these points stated clearly, they went ashore to inquire about [~~the~~] forwarding [~~of~~] the copies, promising to return in an hour or so. During this long confabulation, I tried to get some information of a general nature, but they were rather skittish, refusing to tell by pleading ignorance even of the town north of the point, of the name of the opposite town across the bay, and such like matters.

It was 4 o'clock before the trio came aboard, and then to declare decidedly that they had all along understood that the originals were to be received, and that an envoy had come, whose credentials should be presented as evidence of his true character beforehand. The principal points were then stated in writing, — that the Commodore would deliver the originals & copies together at any designated place on shore, that he would return for an answer, that he must see the credentials of his host, that he should come ashore with a suitable escort, and that no conference respecting the contents of the papers was expected when they were presented, but merely a ceremonial visit. The constant fear on their part evidently was that we meant more than we said, and had designs <u>sub</u> <u>rosâ</u> [Latin=under the rose; privately]; they were referred to the letters as containing all we came for, and told that these must be answered or consulted; hints were also given of our going up the bay.

At our request, Tatsnoski showed his swords to the company. The scabbard of one was covered with a white-brown speckled fish skin, which he said was brought from China; perhaps it is from Manchuria; it was smooth and nicely covered the wooden sheath. The other was covered with hair beautifully lackered & wound around. The blade was rather sharp, quite plain, and bright, but not superior to our's, at least judging by the looks; two gold dragons ornamented the ends of the hilt, wh[ic]h was long, for two hands, & covered with knotted silk. These swords are worn in a most inconvenient way for our custom of sitting in chairs, but not for their usage of squatting. The prices were 20 or 30 taels for the small & large one's.

After all points were explained, they requested to see the engine, and were taken thro' the ship. The size of the machinery seemed to gratify & amaze them, and every principle of propulsion was explained as well as the time allowed. Yezaimon on seeing coal, said that Japan produced it in many places as Firado I., Awa in Sikokf, & Yamatto, besides others; its uses he knew, and was far from making himself foolish as the man did who got a piece from the "Preble"[20] at Nagasaki. The size of the furnaces and the complicated nature of the machinery, drew their wonderful gaze. The guns, muskets, & all the arrangements of the ship[s], the small proportion of the sick out of the 300 souls in her, were all informed them, and they observed everything; a daguerreotype pleased them much, they having previously heard of the name. The survey of such a steamer evidently gratified a reasonable curiosity.

From the interpreter Tokoshiuro 立石(タテイシ)得(トク)十(シユウ)郎(ロー)光(ミツ)定(サダ), I learned that the nengo [年号 : year] of the present cubo [公方 : Tycoon] is Kayei 嘉永 , & this his sixth year; his predecessor was Ohoka 弘 化 and before him was Tenpo, 天 保 , who ruled when we were here in 1837. These monarchs do not reign so long as their brother emperors at Peking, & I suspect have less power and influence in the state; if the story be true that they are required to resign whenever they are in the minority with the state council on public questions, it is no wonder their reigns are brief. He also gave me the official title of Yezaimon 浦 (ウ ラ) 賀 (ガ) 騎 (キ) 士 (ス) 長 (チ ヨ) which is literally, the "Uraga riding elder scholar", but what this means I do not know; his subordinate, who came aboard the ship first, named Nagazhima Saboroske 中 (ナ ガ) 島 (ジ マ) 三郎 (サ ブ ロ) 助 (ス ケ) is styled 浦賀 (ウ ラ ガ) 騎隊 (キ ス イ) the "Uraga Rider of a battalion", which is alike obscure; his duties seemed to include those of port warden among others.

July 13th

The officials did not reach the ship till 4 o'clock today, alledging the non-arrival of the envoy from Yedo until late in the day. Yezaimon bro[ugh]'t the credetials of this commissioner and a translation in Dutch, but no copy in Japanese or Chinese, so that it is impossible to verify the certainty of this translation, tho' I do not suppose any deception is to be feared. He was rather sensitive when I came up to him to see the paper, and stipulated beforehand that it was not to go out of his hands; the seal was a small round one in the seal character, and was stamped once in halves by folding the paper over so as to bisect the impression; the paper was common, and the whole was carried in a case in the bosom of the dress. I suggested the propriety of having a copy in the original, but it was overruled. Many points respecting the interview were settled; the place was Hori-yama around a point below Uraga, and the size and composition of the escort was inquired into. One difficulty on their part came out, which was the trouble of seating so many foreigners in a country where the people all squatted, but we told them it was unimportant, and the Commodore would take the same accommodation the envoy had.

Three of their attendants walked over the ship while these three were in the cabin, and expressed their thanks afterwards at the sight, which was one they had perhaps wished for, since they had often come off before; the oldest of them wished to know if the women in U.S. were white; and then where I had learned his language; in explaining the last I told him there were many Japanese sailors abroad. The way in which this man talked gave me the impression that freer intercourse with foreigners w[d.] please many thousands of people in Japan, if the restriction now existing is divested of all danger, and [that] people can do as they like to their visitors.

The suspicious character of the officials seemed to show itself plainly today, but their inquiries may have been forced upon them, and they obliged to ask so many questions to satisfy their superiors, who had not had their opportunities.

July 14th

The squadron was full of bustle this morning, getting arms burnished, boats ready, steam up, men dressed, and making all the preparations necessary to go ashore, and be prepared for any alternative. About $7^1/_2$ o'clock, the steamers were under weigh, and soon opened the beach around the point, and disclosed the preparations made to receive the letters from Pres. Fillmore. The officials, in their boats, were lying off the "Susquehanna" waiting to see the flag hoisted, and about the time our anchor was down, they were alongside. There were two boats carrying six officials dressed in full costume, who when seated on deck presented a most singularly grotesque & piebald appearance blended with a certain degree of richness from the gay colors they wore. The 2nd officer was a conspicuous member of this party, he not having been aboard before since the first day; and his dark face & sharp features contrasting with his yellow robe, and his black socks, having bare legs, and short trowsers all showing out from the overalls of his uniform made him rather an attractive object. I cannot describe the dresses of these men minutely, but the effect was not unpleasant, tho' in most of them no harmony of colors was aimed at in the uniforms. They all seemed to be in good spirits, and amused themselves looking at the officers in their uniforms, & other things.

By 10 o'clock, the boats had left the steamer, and under the lead of the natives were pretty much landed before 11 o'clock on the beach at Hori-hama 九里濱 or クリハマ, opposite the shed erected for our reception, & surrounded with striped curtains; Com. Perry left under a salute & found the escort ready when he landed to conduct him to the house prepared for his audience. There were 15 boats in all, containing about 300 people, say 112 marines, 40 musicians, 40 officers, and a hundred or more sailors. Every one was armed with a sword, a pistol or a musket, & most of the fire-arms were loaded; I borrowed a coat & sword so as to appear like the rest, but my uniform would hardly bear inspection or classification. A jetty had been made of bundles of straw covered with sand, and facilitated the landing very greatly. The precaution of bringing down the two steamers to cover the place of meeting made it easy to land from them without exposure to the sun; the bay near shore was deep but full of seaweed growing in long leaves to near the surface, and doubtless full of marine productions.

The place appointed for receiving these letters was a hut set up on the beach, having two small ones behind it, the whole inclosed by white & blue striped curtains hanging from poles; a screen was in front, concealing the front of the room, and a large opening at each end of it, between that and the side curtains, which were prolonged along the beach on each side hand for nearly half a mile. The village was in the south of the cove, near the course from whence the "Morrison" was fired at, a poor hamlet of 200 thatched huts, mostly concealed from our view by the curtains and the crowd. The hills rose behind, partly cultivated and looking exceedingly fresh and green, inviting us in vain to explore their slopes, for the ridiculous laws interfere to prevent our trespassing on them. Truly, laws which prevent such things must have been brought about by a hard & dear experience, for it is against nature thus to prohibit intercourse between man & man.

The Japanese had placed a row of armed boats near the ends of the curtains, and detachments of troops were stationed before the curtains in close array, standing to their arms, their pennons flying from the curtains, and gradually bending down to meet the boats at each end. Some of these troops were dressed in dirty white in a manner similar to the troops in Egypt, with full breeches & tight stockings; others resembled Chinese troops, and many were in a tightly-fitting habit. Horsemen were placed behind one or two curtains, who wore brass cuirasses and metallic helmets or something like it. Their horses were large animals, far beyond the Chinese beasts I have seen in size, and looking like another race than the little Lewchewan ponies. All these troops, (numbering about 5000 men as one of the Japanese told me,) maintained the utmost order, nor did the populace intrude beyond the guard. A few miserable field pieces stood in front, not over $4^{lb.}$ [=4 pounder] or 5^{lb}-er [=5 pounder] I should think; many files had muskets with bayonets, others had spears, and most I could not see. Crowds of women were noticed by some near the markee [=French, marquee] but I suspect they were not numerous. Altogether, the Japanese had taken great pains to receive us in style, while each side had provided against surprise from the other, and prepared against every contingency.

As soon as Com$^{re.}$ Perry landed, all fell into procession; Capt. Buchanan, who was the first man ashore, had arranged all in their places, so that no hindrance took place. The marines, headed by Major Leilen led off, he going ahead with a drawn sword; then half of the sailors, with one band playing between the[~~m~~] two parties. Two tall blacks heavily armed, supported as tall a standard bearer, carrying a Commodore's pennant, went next before two boys carrying the President's letter & the Full Powers in their boxes covered with red baize. The Commodore, supported by Capt. Adams & Lt. Contee, each wearing chapeaux, then advanced; the interpreters & secretary came next, succeeded by Capt. Buchanan & the gay-appearing file of officers, whose epaulettes, buttons &c. shone brightly in the sun. A file of sailors & the band, with marines under Captain Slack, finished this remarkable escort. The escort of Von Resanoff at Nagasaki of seven men was denied a landing until they had been stripped of almost everything belonging to a guard of honor[21]; here, 50 years after, a strongly armed escort of 300 Americans does honor to their President's letter at the other end of the empire, the Japanese being anxious only to know the size and arrangement of what they felt themselves powerless to resist. There were fully a thousand charges of ball in the escort, besides the contents of the cartridge-boxes. Any treachery on their part would have met a serious revenge.

On reaching the front of the markee, the two envoys were seen seated on camp-stools, on the left side of a room, 20 ft. square or so, matted & covered with red felt; four camp stools were ranged on the right side, and a red lackered box between them. The chief envoy, 戸田（トダ）伊豆（イヅノ）守（カミ）Toda, Idzu no kami, (Toda, prince of Idzu) and his coadjustor, 井戸（イド）石見（イワミ）守（ノカミ）Ido, Iwami no kami (Ido, prince of Iwami) rose as the Commodore entered, & the two parties made [~~a~~] slight bows to each other. The boys laid the boxes on the floor, & the two blacks came in to open them. They were taken out and opened upon the lackered-box, and the packet containing the copies and translations presented by Mr. Contee. Tatsnoske and Yezaimon were both on the floor, and the former commenced the interview by asking if the letters were ready to be delivered. When he made known the reply, he put his head nearly to the floor in speaking to Yezaimon, who was on his knees informed the envoy in a whisper. The receipt for them in Dutch & Japanese was then delivered to Mr. Portman; and the originals themselves opened out in the boxes as they lay. Soon after, Com. Perry said that in two or three days he intended to leave for Lewchew & [~~Japan~~] China, and would take any letters, &c., for the envoys. This produced no acknowledgement on their part; and he then added that there was a revolution in

China by insurgents, who had taken Nanking & Amoy, and wished to introduce a new religion. "It will be better not to talk about revolutions at this time," was the significant reply; and proper one, too, for I thought it very mal-apropos [=untimely] to bring in such a topic. Yet one might regard it with interest as ominous of the important changes which might now be coming on the Japanese, & of which this interview was a good commencement.

Conversation being thus stopped, & no signs of any refreshment appearing, there was nothing else to do than to go. The contrast between its interlocutors was very striking. In the front was a group of foreign officers and behind them the picturesque-looking shaven-pated Japanese in relief against the checked screen; on the left a row of full-dressed officers, with swords, spaulettes &c, all in full lustre; on the right the two envoys and a secretary, with two more plainly dressed men on their knees between the two rows. To describe the robes of these two envoys is difficult. The upper mantilla was a slate col[ore]'d brocade kind of silk, made stiff at the shoulders so as to stick out squarely; the girdle a brown color, & the overall trowsers of purplish silk; the swords were not very rich-looking. The coat of arms was conspicuous on the sleeves, & some of the under-garments appearing, gave a peculiarly harlequin-like too to his dress, to which the other envoy was accordant. They were immovable & never stirred or hardly spoke during the interview; and one who tarried a little as we came out, said that they relaxed in their stiffness as soon as we had gone, apparently glad that all was over. I got the impression that the two high men had pursed themselves up to an attitude, and had taken on this demure look as part of it, but others looked on it as a subdued manner, as if afraid. The reëmbarkation took place gradually, no one being in much of a hurry, & I began to talk to the people, and invited two of them on board to see the steamer & a revolver. One man wished to know if the women in America were white; another, how he could learn strategy, to whh I replied, "only by your going abroad or letting us come here." I asked him why there was no music, to which he answered that it was very poor. Considerable curiosity was manifested in comparing swords, & some exchanges were proposed; altogether this part of the interview was far the pleasantest to both parties, and I suspect the Japanese were sorry to see the show end so soon. Many picked up shells & pebbles to remember the spot, & by 1 o'clock every body was back to his place.

Two boats full of people came alongside soon after, and stayed on board while we steamed back to Uraga. Yezaimon especially, took much interest in seeing the working of such stupendous machinery, & inquiring into the manner of turning the wheels. All was made plain as we could explain it, tho' I fear the ideas were very crudely expressed, for I did not know their language well enough, & Portman seemed not to know the machine well enough.

[a business card in Japanese: 浦賀与力應接掛 中島三郎助] Yoriki Oösetz' gakari Card of N. Saboroske]

[a business card in Japanese: 筵倉相太郎] Signature of a military officer at Uraga

[a seal in Japanese] Copy of seal of Prince of Iwami, Ido.

[a seal or signature in Japanese] Copy of seal or Signature of prince of Idzu, Toda.

Their sec^{y} was 辻（ツジ）茂（モ）右（ウ）衛（エ）門（モン）Tsuzhimo Uyemon. One of our visitors was the military commander of Uraga, an open faced pleasant man, who wished to learn something of tactics, and the construction of revolvers. One of the pistols was fired off by Capt. Buchanan to gratify him & Saboroske, and they had many measurements to take of the cannon on deck; the latter greatly [am]used us by going thro' the manual with a gun he took off [from a pistol-] stand, his face pursed up as if he was a valiant hero. This man is altogether the most forward, disagreeable officer we have had on board, and shows badly among the generally polite men we have hitherto had, prying round into everything and turning over all he saw. At our request the party remained on board while we steamed up to Uraga, and then bid us good-bye, having made themselves conspicuous in every part of the ship by their particolored dresses. Some refreshments were given them in the cabin, and they went off in good humor.

The receipt given by the two envoys was to this purport: "According to Japanese law it is illegal for any paper to be received from foreign countries except at Nagasaki, but as the Commodore has taken much trouble to bring the letter of the President here, it is notwithstanding received. No conversation can be allowed, and as soon as the documents and the copy are handed over, you will leave." The Japanese original is written on very thick paper, made from the mulberry (Broussonnetia): the last sentence of it intimated they were to make sail immediately.

The four ships now stood up the bay, and anchored about where the "Mississippi" had sounded, some 12 miles above Uraga. Erelong, Yezaimon appeared alongside, looking sour enough at this his third visit to the "Susquehanna" today. His object was soon explained, and we endeavored to ease his mind in respect to surveying the harbor, telling him that [we] had told him we were not going to sail immediately, but [move] about the bay, & seek a better anchorage than that off Uraga for placing our ships next year.

The extent of the time we should stay could not be stated, but not likely to exceed four days; we would not land, nor would there be any trouble of the Japanese made none, for our boats were strictly ordered to abstain from their's. I think he himself was satisfied of our intentions, but his superiors were probably alarmed at the risk, and sent him to do what he could to prevent further progress. The interview was rather tedious from its being a struggle, and I suspect the interlocutors were all pleased when it was over. Others from the boat came on board, & walked through the ship, and I wish there were more who could have seen her. At this visit and the one earlier in the afternoon, many things were shown our visitors, such as engravings, daguerreotypes, & curiosities of various sorts which tended to relieve the monotony of the visit, as well as instruct them a little. I have now learned more fluency by my practice, & did considerable side talking.

At eventide we were left alone, & thus closed this eventful day, one which will be a day to be noted in the history of Japan, one on which the key was put into the lock, and a beginning made to do away with the long seclusion of this nation, for I incline to think that the reception of such a letter in such a public manner involves its consideration if not its acceptance[,] at least the prestige of determined seclusion on her part is gone, after the meeting at Gori-hama.

July 15th

The "Saratoga" & "Plymoth" came up today from the anchorage off Uraga in Lat. 35°15' N, Long 139°49'

E. to join the two steamers at the "American Anchorage" in 35°23' N, 139°41' E, off a thinly inhabited coast. The shores were much more wooded here than off Uraga, and steeper. North of us on a low projecting point were seen many pennons and increasing crowds of people, perhaps many of them soldiers brought or attracted from Kanagawa & the interjacent country to see us. No signs [~~n~~]or words could attract any of the numerous boats to draw even within fair speaking distance. The surveying boats went up in the morning almost out of sight, & in the afternoon, the Commodore [~~went~~] [~~up~~] proceeded in the Mississippi over the same & some new ground. The town of [~~Kanagawa~~] Kawasaki stretched along the north bank of the [~~river~~] Taba gawa, a well placed & populous town. We tho't at the time that this was Kanagawa, but by Siebold't map that town lies south, & on no stream[,] a little island[,] not far from our "Am$^{n.}$" Anchorage", and the people who come on board seem so chary of telling the name of a single place, that one cannot feel confident they tell it right when they do give it. There were many vessels entering, & more at anchor in the river, which seemed a wide stream near the town. Nothing of Yedo could be distinguished, but a long serried row of masts seemed to indicate the position of Shinagawa the suburb port of the capital. A singular shaped structure in the bay, seemed to limit the vessel's going-up-track on the east; Sam Patch called it Boögi, and describes it as a tree on an islet. It looked like a steamer coming end on, with an enormous smoke pipe, or a round-house with a tower rising from the midst; he said it had nothing to do with ships, and in fact knew almost nothing about it, except its existence, & that Yedo was 3 or 4 ri northwesterly from it. The land east of this was too low to see more than the trees & hills, but no signs of islands appeared from our ship anywhere, and the land rose on the NE & E shores. We estimated ourselves to be 10 miles from Yedo, & turned about at even.g in 17 fathoms, pleased with having had a look at Kawasaki and as far ahead of it as we could see. The shores were well wooded, but the population did not apparently increase as we neared the city, and we were obliged to turn back without a sight of the goal.

On returning to the "Susquehanna", we learned that Yezaimon had come alongside with some presents, which were declined until the Commodore could be seen. He looked disappointed, but was told to come again in the morning as soon as the flag was up. A surveying party also returned at evening to report. It has penetrated up a creek, where some intercourse with the people of a village had been held from the boats, the whole population, men, women & children, running down to see the foreigners from the beach, & showing much pleasure at the chance. Some water & green peaches were procured from them, and all that was wanted was ability to understand each other. There was some motions made of cutting throats, but no one seemed to regard them otherwise than gestures, and the two parties separated, much gratified with their unexpected interview. The country along the creek and coast was pretty, but not much settled. It is truly a disappointment to lie off so inviting a country day after day, and be obliged to only spy it thro' glasses, & guess what this & that thing is. Wait till we come again!

July 16th Off Saru sima

We came down to this beautiful islet of Saru sima early this morning; it lies about half way between Uraga and the "Amn. Anchorage", less than half a mile from the shore, and is perhaps 200 or 225 ft. high, prettily wooded, and defended by three forts, made of earth embankments with wide portholes, the walls of these embankments were grassy, and the scrap of the hill behind being likewise grassy, they were almost masked batteries. Few places along the bay have been better chosen than this islet for defending the passage, or for a pleasant residence for troops. The banks along the main land were singularly cultivated in alternate stripes of clearings & copses, giving it a striped look, especially near the village of Otsu.

Almost before we had anchored, Yezaimon came pulling alongside, bringing the presents; the interpreters came in two boats, and showed us a mem$^{o.}$ in Dutch, to the effect that the letter of the President sent thro' the Dutch at Nagasaki had been received, & that probably our present letter would be favorably regarded by the Council, but that it rather worked against us (by what manner was not intimated) to be cruising about the Bay and examining it as we did. This paper received no notice, being merely a mem$^{o.}$ such as we had given them, and yet its contents were evidently directly pointed to attain our departure as soon as possible, by holding out the hope of attaining our end. It is not unlikely, therefore, if we could remain in the Bay a month, showing the ships here & there, that the great ends of the mission might be obtained now in order to avoid a second visit.

Yezaimon & his suite took breakfast with Capts. Buchanan & Adams, and behaved themselves very properly. The presents in return for their's were ready about 9 o'clock, consisting of 1 box of tea, 3 engravings of steamers & a house, 3 History of U.S., 20 ps. of coarse cotton, bale of drillings, a loaf of sugar, box of champagne & demijohn of whisky; they declining to receive the 3 swords. Their's were 5 pieces of brocade, 40 bamboo fans, 50 tobacco pipes & 50 lacquered cups, which were described merely as tokens of remembrance, and they wished us to receive them as personal favors. Considerable discussion ensued on this point; they wished to leave their's on board & ask permission to take our's in the afternoon, or to send ashore to ask their superiors, but no alternative could be allowed: they must either take our's, or carry their own back again, and we had began to put them up to be replaced in their boat, when they agreed to the least serious alternative for them, & went off with the Commodore's presents & list, taking a few other mementoes from us who had had most of the conferences with them, such as corns, soap, pictures &c. I have no doubt they kept the whole themselves, concealing the transaction (as an exchange on equal terms) from their superiors.

During the day a survey of this part of the bay was completed, the two sloops came down to the spot; and when, in the afternoon, Yezaimon came off to bring a parting douceur [French=gift] of fowls & eggs, we were able to reässure him that the squadron would sail in the morning, as we had promised him when at breakfast. His assortment of fowls was rather a pretty collection of bantam & other kinds; & he made no objection to receiving a box of seeds, two cakes, bottles of cologne, cherry cordial, maraschino, & some cakes of soap, besides a good potation of punch & champagne under his girdle. He was in very good humor with everybody, and left us, with all his retinue, about 5 o'clock, having visited the ship every day since he first came off a week ago this morning. In all his conduct, he has shown great propriety, apparently never getting out of humor, and exhibiting no hauteur or acerbity towards his inferiors; listening to whatever was told him with all courtesy, whatever was its purport.

At this and other interviews, we endeavored to please our visitors by showing them pictures of various things, daguerreotypes, & other little articles. I showed the map of Yedo I had, and they pointed out some places on it, saying that the city had very much increased in the 86 years since the map was drawn; they asked no questions relating to it, and were disinclined to answer many, for geography seems to be a delicate subject whenever alluded to in any way. On their part no general questions were asked, so far as I now remember, at any interview, except the names of those whom they met in conference; nor did they exhibit a single article of curiosity or show the least willingness to exchange any thing as mementoes, except a fan which Yezaimon & I passed.

They cannot, I should think, conceal from themselves, that during the last week their government has let down the principle of seclusion it has hitherto maintained in refusing all intercourse with foreign nations, except the pent-up, despicable communications held with the Dutch & Chinese at Nagasaki, which must have tended to exalt their own importance, & nourish their conceit in a great degree. Let any one read Langsdorff's account of the treatment of the Dutch at Nagasaki, and note their complying demeanor to all the insolence of the officials, and his detail of the indignities Resanoff was obliged to submit to from the same men when he was there in a half crippled condition in a leaky ship, and was put off by the most trifling impertinent excuses, and compare them with the incidents here given, must see that we have made a very different impression upon the government, and led the chief rulers to adopt entirely a different course, whether from fear or deliberative purpose, or whatever other reason.[22] I pray God most humbly to order all future events so that the seclusion hitherto maintained may be removed without any collision, & open the way for the introduction of this people to their fellowmen, and their gradual elevation in science, arts & true religion.

July 17th

We got under weigh this morning, and each steamer taking a sloop in tow, passed out of the bay at the rate of 9 knots, in a calm, showing most plainly the power of steam to the thousands who watched us. The houses at Gori-hama were still standing, and the pennons [~~were~~] fluttering at most of the forts, with a number of the curtains still stretched out, but not many troops appeared. At the par[t] near Cape Sagami, fully a thousand boats were seen, all of them small ones & without sails, each containing 6 to 10 people, apparently abroad for no other object than to see the ships depart. To a maritime people, the contrast between their weak junks & slight shallops, and these powerful vessels, must have made a deep impression.

During the day, we passed down among the islands off the Bay, and noted three not laid down in our charts, which were immediately labeled by our officers after the three ships, the Plymouth having already been accommodated with a rock. These islets seemed uninhabited, but this conclusion may be erroneous. Vulcan I. exhibited no smoke, & looked invitingly green, so that its fires may have gone out in late years.

Monday, July 25th

During the last week, we have been making slow

progress, chiefly owing to bad weather, which came on within a few hours of the change of the full moon. "Saratoga" was in tow all day monday [*sic*], but her two chief officers were called on board to receive orders, and when they went back, took my two Chinese to land in Shanghai on board with them. The two ships let loose their hawsers Tuesday afternoon, and next morning were just in sight ahead. Wednesday we had a strong NE. wind, & Thursday it had increased so that we lay to[~~o~~], heading SE. for 24 hrs., and then NE. for most part of Friday, the sea being very cross and high, indicating more severity of wind than we had not far from us. The yards & topmasts were sent down, guns lashed, & steam reduced, whereby no damage was sustained. The reason for all this caution was the desire to see & examine the O-shima Is. which lie NNE. of Lewchew, after the sea & wind abated; but by Friday noon, it was decided to go straight to Napa, and defer their inspection till a more fitting time and pleasanter weather. The wind remained steadily at the east, and we made one p$^{t.}$ of Barrow Bay yesterday morning, and expected to get in to port in the evening; but as it thickened up towards night, the commodore stood off when within only 6 or 8 miles of it, and bore away to the S. & W. We got up steam early this morning, and after running about 25 miles cast anchor in Napa Harbor, the expenditure of coal for this cautious movement being about $500, and Perry almost the only one in favor of it. However, none under him had the responsibility. On anchoring, we found that the "Supply" had experienced bad weather on Saturday & Sunday morning; and the swell from the S.W. & S. indicated more wind in that direction, so that we may have escaped the worst of the gale by laying to as we did on Thursday, for it has plainly traveled from NE. to SW. The Plymouth's logbook will assist in tracing its direction & force.

As usual, I was so seasick as to be unable to do any work, and could get little comfort from Mr. Taylor, who was if anything rather worse. This penalty is now over, however, and I am thankful we are safely back without any mishap to crews or ships. Many are disappointed in not finding the "Powhatan" in port, but I shall be glad to see the "Plymouth" showing herself off the harbor in good condition, and the "Captrice" following her in like order.

July 26th

The Mayor of Napa has been wise enough to resign his office within a day or two, and his successor Máu Yuh-lin 毛玉麟 sent his cards off yesterday evening to the Commodore & Capt. Lee, and the messengers tried to learn something of our visit to Japan & its results, but I turned them off by promising to return their visit tomorrow, & telling them then. This morning, accordingly, Capt. Adams & I waited on the new mayor, a far inferior official in his bearing and energy to the former, and apparently older; the other man, I suspect, has had enough of interviews and dinners, & retires to safe retirement, before he embroils himself. Several points were submitted to the mayor this morning, which he was unprepared to answer directly, and did not wish to [~~do~~] [~~so~~] at all. We thanked the govt. for erecting a tombstone over the grave of Pons, and wished to learn the cost in order to repay the same; but as they declined to mention it, we told them that they need not put up one over the man buried yesterday from the Mississippi. The rental of the house at Tumai for a year was demanded, in order to pay it; but they alleged that it was a temple, & no rent was charged for occupying it. Room near it was requested on which to get a storehouse built to put coal in, which was to be built by the gov[t.] and rent paid for it, or else we would have it erected by native workmen. It was demanded of them that the spies who followed officers whenever they walked abroad should be removed by their superiors, and fair warning was given that if any collison took place, and injury rec[d.] by these tag-tails, it would be their fault. Two months had shown that we did them no harm, and we did not wish to have the women & children running from us because these underlings were in sight. We wished also to buy articles, & the commodore wished particularly to get a great variety of articles, silks, cottons, lacquered-ware, china-ware, & other products, to put in a museum in Washington. The commodore also desired to have an interview with the Regent, to discuss these points, & it was agreed that I should come tomorrow & learn the time & place for the meeting, as the Regent should appoint. These "heads of discourse" were all written in their presence, & they were advised to deliberate on them satisfactorily to us. We remained there a long time, for we could get no definite answer to these requirements, and indeed hardly expected it. I admonished Ichirazichi about the spies, & told him that the officers might carry pistols, and hurt some of them if they persisted in tagging after & constantly interfering wherever we went; I hope the hint will be passed on to the spies themselves, who, after all, are only peaceably doing what[~~ever~~] they have been ordered, & should not suffer. The whole interview was less engaging than previous ones here from the less pleasant manner of the Mayor, who took no pains to show the least interest in what was told him. Perhaps this qualification has been his recommendation to the post at this time.

July 27th

A long document addressed to the Regent was drawn up this morning and carried ashore by Lt. Contee & me to deliver to the Mayor. We reached the kung-kwan about noon, where we found a smaller coterie of officials than were present yesterday. The paper was a threatening expostulation at being treated so unfriendly, disallowed access to the markets & shops, followed by spies into every corner and lane, who prevented all intercourse, and held at arm's length in a way we would not admit was right, nor submit to; & if a change was not made, means would be found to bring it about, on a return to Napa. The Mayor declined opening the envelope, & promised to forward it to Shui. The place we were in, and 2 o'clock tomorrow, were appointed by the mayor as the time & place for the Commdore to see the Regent, altho' it was tried to get 12 o'clock as the time. No answer could be got out of them with respect to the demands made yesterday, but answers were promised at the meeting.

Ichirazichi then proposed some questions respecting our visit to Yedo, but after saying that there had been no fighting & we had gone ashore, I referred him to the morrow's meeting for all particulars. He asked if the ship which came in this morning was the Plymouth, and if the st[r.] Mississippi was named from the state of Mississippi, & how many stars we had now in our flag? From these questions, I saw that he had been reading the History of the U.S. given him, & then I asked him some more names, & told him he must go to America next year, & see for himself. He demurred on account of the length of the voyage, &c., but perhaps the idea is not unpleasant to him.

After munching melons & cakes, sipping tea, talking &

scolding, for an hour, we left, & made a crooked road back thro’ the town to Junk harbor, going thro’ the dirty pork market & along the creek, till we reached the end of the pier. The view of the surf as it came rolling in over the reef was fine. When the boat came for us, we took a stroll thro’ a village across the harbor & a pull up to the watering-place. The southern bank of the river is very prettily terraced, & everywhere under constant cultivation, showing that much of the supplies of the town are brought from this region. How much this pull reminded me of the attempt we made to see the town of Napa from the Morrison’s gig by pulling up to the top of the river! Every point and turn seemed to be familiar, tho’ it is probable that what I saw then has all passed out of mind. In the evening, [~~of~~] a party returned from a visit to an old castle lying southeast of Napa, which was described as being an aggregate of large houses and walls, apparently very old and ruinous, & not so strongly built as the one [~~of~~] at Shui.

July 28th

At 2 o’clock, Com. Perry and suite, 17 in all, left the ships, to pay a visit to the Regent at Napa kung-kwan, altho’ we had just learned from a messenger sent off to the ship that he had been ready at noon, & was waiting for us; why he was unwilling to agree to have the meeting at noon when requested, was not easy to understand. We landed near Capstan Pt., and after waiting a while for other boats, & being joined by Dr. Bettleheim at the Comre.’s invitation, went directly across to the main street to the kung-kwan, where the Mayor met us outside of the gate, & the Regent inside; the latter took Perry’s arm & led him to his seat, and waited till all had got their places before sitting.

Compliments having been passed, the Commodore said that he wished to speak upon business before easting, & that he hoped the Regent had deliberated upon the points offered for his consideration two days before, & had an answer prepared. The Americans were people of few words & many acts, and wished now to come to a fair understanding, as they meant what they said & no more: that they had come to Lewchew in a friendly spirit, & expected to be received in the same way they were in China. The Regent replied that an answer would be ready, and invited his guests to partake of the eatables spread out before them. He maintained the same impassible, fixed position & look, as when on board ship, constantly glancing his eyes about; his coadjutor indicated little interest in anything. After a little, questions were propounded respecting our doings in Japan, when the Commodore told him that we had visited the Bay of Yedo, had been received in a friendly manner, had gone ashore with about 400 persons to meet the Princes of Idzu & Iwami, when over 5,000 spectators were assembled, of whom 1,500 were soldiers, had exchanged presents, & gone within 30 li. of Yedo, anchoring and sounding in such parts of the Bay as we pleased; and lastly, that we were going back there next year. There were more questions ready, but as they were told all the important points, it was deemed best to bring them back to the subject in hand, & have them answer our questions first, before talking further upon Japan.

We went on eating a while, some 6 or 8 courses of stewed dishes, following slowly as their forerunners disappeared, when the Commodore called up the Regent's reply; a little before this, Ichirazichi being aside, I asked Bettelheim to tell the Regent t[ha]t the Commodre thought it would be well to send two of the waiters to U.S. to spend a year or two in learning our language, but the official would not hear the remark until it had gone thro' the lips of one of té-fú, greatly to Dr. B.'s amusement & perhaps annoyance. The Regent seemed to have been starched up for the occasion, & his position was as definite as an orderly serjeant's.

At last the paper came, & the Regent took it, left his seat, & went in front of the Com[re.], & politely handed it to him; he was requested to be seated again till it could be read, & Perry then took his seat. It began by recapitulating all the items given the Mayor on Tuesday, word for word as I had written them, as they had been reported by that functionary to him (the Regent). To the proposal to pay rent, it was urged that the priests who had temporarily vacated the house now occupied by the squadron, could not rent their lodging & find another; and therefore it was inconvenient to receive the rental or have it occupied. The demand to have a coal dépot near it was turned off, by a repetition of their being poor, and that if such a place was erected, they would be overwhelmed with care & trouble in looking after it; for Bettleheim had already remained here some years, & given them much trouble, & now if we came too, building & lodging, their poor country could not stand it. In regard to buying & selling, they had nothing to do with the proceedings of shopkeepers & marketmen, who opened & shut their shops, & sold or retained their wares, just as they pleased; but added, that their own productions were exceedingly few, & manufactures contemptible, all they had coming from China & Japan, of which only a few lots of the silks, chinaware, lacqd. ware & cloths came from those countries. The last article, concerning the spies following us, was plainly granted, as we had expressed our dislike at them, and said that they were no assistance, protection, or use to us when going about. Probably the frequent recurrence to this topic in our interviews, the paper handed in yesterday, & the consciousness that a collision might ensue in some bye-path, led them to adopt this resolution. It closed with an earnest petition that the Commodore would receive this reply, and have compassion on them.

As soon as I had read it, he ordered it to be returned to the Regent as being so different from what he expected that he gave it back for further consideration. The poor man came forward again, & would have made a kotau, if I had not stopped him. The position would not be rec[d.], & must be discussed more favorably to us by tomorrow noon, & brought on board, or else the points would be referred to the Prince at Shui, & we should go there with a large party & wait till we got an answer. As to the dépot, if they would not build it, or allow us to do so by employing natives, the materials should be brought & the house erected. Much time would not elapse before the authorities would feel it was best for them to agree to our wishes; for in China we had no trouble in getting such facilities, and there was no danger in their furnishing them here.

The Commodore left in a few moments, and perhaps nothing further could have been said with any avail. It was a struggle between weakness & might, & the islanders must go to the wall; it was as well planned on their part as possible, and they were doubtless disappointed at the result. Taking the question in all its bearings, I really pity them much, for the rulers here form an oppressive oligarchy, and ride the people to extremes, even to the non-fruition of their own wishes and gain, and the continual impoverishment and degradation of the latter. The scene had some tragic features, perhaps many more than appeared, and was in every view a reality to the natives, however much of a dramatic character was mixed with it in our eyes. The seclusion of these islanders must give way, and if nothing worse comes than the granting of these demands, they will certainly be the gainers, and their policy will have time to adapt itself to the new influences now felt.

July 29th

About 11 o'clock the querulous mayor & Ichirazichi came off with two or three others, the old man being evidently discomposed by his trip and the heat; excuses were made for the Regent, who may well have been excused from the retraction of his yesterday's petition. The interpreter began by asserting the propriety of the paper presented, but the chief point of refusing the dépot was peremptorily overruled by our saying that we should build it oursevels if they did not, but it must be close by the landing as the house was too far from the boats. Excuses were made, then, that tyfoons wd. destroy so exposed a house, or thieves [~~would~~] pilfer coal lying so remote from careful officials (& here a sad picture of the morals of the people in regard to meum & tuum [Latin: meum et tuum=mine and yours] was dreary) or laborers would be scarce to erect it, and lastly, that they would alter the house adjoining the main building in the yard for this purpose. All these doublings were overruled, and the previous question was carried by our appointing a meeting on the ground at 2 o'clock to stake out the limits; and I have little doubt but that they came with this ultimatum from their superiors.

The purchasing of articles & provisions was a mixed question, for we already got the latter, (tho' I scolded them for their non-fulfilment of orders,) and I think could not get them with less trouble to ourselves; it is out of the question to have the ships supplied with boats coming alongside as in China, for a long time to come, and who is to go to the dirty markets and pick up eggs & chickens? The plan was now pursued is perhaps more expensive to us, and profitable to the officials, who are beginning to see the benefits of such a demand, and these two reasons will combine to keep the present way in operation. It was, however, agreed that on Monday an assortment of every article should be spread out in the Napa kung-kwan, where the Commodore would go and purchase; particular directions were given as to the assortment & quantity of articles to be brought, but I have great doubts as to the result of this bazaar.

Thus the two main points were conceded, and the interview ended amicably enough, as far as appearances could indicate, drinking & eating meanwhile, so that at the last, they had pretty well got over their squeamishness. At 2 o'clock Capt. Buchanan & Adams & I were on the spot, but no officials, for whom we sent off [~~by~~] two messengers; meanwhile, we staked out the ground, and found that a sufficiently large spot could be marked out without cutting away any trees of size, or intruding on any useful spot. Three o'clock passed away, and they went [~~aboard~~] ashore, leaving me to meet the authorities, on whom no gentle words were laid, for their tardiness. They came soon after the boat shoved off, & I showed them the place; it was much larger than they had rec[d.] the idea of from our description, and I was myself unable to do more than refer them to the stakes and marks, which were to be the limits; it was much larger than I had supposed would be wanted, and told them I would speak to the Commodore. They were told that they must clear the ground of the rubbish and grass, and a plan would be given them tomorrow. I then went off, for I was hungry enough, in a boat just come, and left them there; by nightfall, the area was nearly cleared of all the shrubs, under the direction of three old graybeards, who superintended operations seated on a mat, & directed the gnomes who flitted about with wisps & twigs which they had gathered up. The scene was very lively, and I thought the natives greatly enjoyed it.

July 30th

My calculations to visit the old castle today were all spoiled by an order to take the plan for the coal dépot ashore & explain it to the builders; we found nobody there on arrival, and were obliged to wait more than three hours before any responsible person came. The details were all clearly understood by means of the diagrams, & the officials required to clear a larger space, & put up the shed as soon as possible, to receive the cargo of the "Caprice", a thatched hut 60 ft. by 35, & about 10 ft. high. I hope those who superintend the job will let the workmen have some of the money received, but I am afraid that they will not get a fair reward: as we drive the officials, they will drive their underlings.

In the evening, during my walk, I found my way into a literary establishment near the bridge, a series of three buildings pleasantly situated behind the stone wall amid a grove of trees; the doorway had a tablet stating that it had been repaired in the 20th year of Kienlung (1755) and by the assistance of the Chinese ambassador here. There were four men writing on small stands in the principal room, who told me that they studied the Nine Classics, but I could not induce them to show me their books. Several tablets were hung up in the room, and the aspect of the whole grounds was retired & scholastic.

The broad way which ran along the edge of this river is one of the thoroughfares of the town, & we watched the passing crowd with attention for a long time. The groups of women & children around some stall or basket, where pattens or pottery, bean curd or pea-sprouts, were sold, engaged our notice by their foolish fear and refusal to have anything to do with us, they would neither take our coppers nor answer our words; the older children shaking their hands in the most seriously comic style. The children are usually in pot-bellied, and remind me of Egyptian children both in color & gait. Sometimes a woman, known by her flowing loose gown to be of a little higher grade, would hurry by us, presenting in her quick step & sidelong glances & turns a growing struggle between fear & curiosity, so that we were sometimes in doubt which would get the mastery; and then would follow a stately official with his girdle largely displayed over his checked dress. Horses overladen & old women carryg. heavy baskets on their heads, frequently went by; and among the crowds we saw few who were maimed or sickly looking. Most of them were thinly clad; [~~and~~] they were generally clean; short and stocky, especially the women, who will not average over 4 ft. 10 in., & may challenge comparison with any other country for coarse features & untidy heads. The men are far their superiors, but it must be remembered that we have not seen the women of the officials, nor any girls reared with care.

Few officials followed any of our officers this evening, from which we may infer that the system of espionage has been pretty much laid aside. Some, who have gone into villages away from Napa have succeeded in getting crowds around them; and further intercourse would doubtless result in our being received everywhere. The sailors in the

Japanese junks have generally showed pleasure at our visits, tho' nothing of any value has been procured there.

We stopped at Dr. Bettleheim's to bid him goodbye, and found others there on a similar errand, more as a mark of respect than goodwill. While his wife has grown in the good opinion of the squadron, he has contrived to get the suspicion or actual dislike of almost everybody. His intrusion into the interview last Thursday was little pleasing to the principal actors, and tends to mix us up with him in the minds of the native authorities. His proceedings have been so anomalous that I am really unable to say what & how much good he is doing, tho' I hope he will come out bright at the last, and his work stand the fire. The council and opinion of a fellow-laborer would do him service & enable his patrons to form a better judgment.

July 31st, [Sunday]

I dined with Capt. Lee & Rev. Mr. Jones today; the Mississippi is a much quieter ship than this on the Sabbath, and to increase the turmoil of washing decks after coaling, most of the men were sent ashore to wash. A dash of rain interrupted the service before Mr. Jones had got thro' the exordium [Latin=preface] of an astronomical discourse he had commenced. Com. Perry seemed rather pleased this morning to report that most of the timbers for the coal dépot were on the ground, and the whole would be done erelong; —all which I suppose is to be laid to the effect of the threat to visit Shui.

Aug. 1st

About 6 o'clock this morning, I was called to go ashore

with Com. Perry to the bazaar opened for our benefit, at the Napa kung kwan. We found a larger assortment than I had expected, and all the finer articles were taken, perhaps in all to the amount of $60; if more time had been allowed, I think we should have had finer pieces brought in from the dealers, and spent double what we did. There was no porcelain, nor many silks, and the whole lot was perhaps not worth over $150, but it will serve as a commencement, and I think the sellers had no cause to complain.

As soon as all returned aboard, the anchor was weighed, & we bid good-bye to Napa, the main demands of Thursday's interview having been all granted. It is doubtful to my mind how much influence the threat of going to Shui & occupying the palace, had in inducing acquiescence, in comparison of the announcement made at the same time and subsequently that we should soon leave if these demands were allowed. Yet the assortment of things this morning showed that the gov't had made known the opportunity to many traders for them to take advantage of, and I hope none of them lost. The stocks were in one or two cases so soon replenished, that the stores could not have been far off; and perhaps even finer ones would gradually have been produced had time allowed. Lacquered bowls & boxes, cotton clothes, silk, & both mixed, hair-pins, sacks, shoes, pipes, fans, coarse pottery, and umbrellas, comprised the list; good prices would have induced them soon to bring more real Japanese lacquered-ware.

We have been at this port about 30 days, & doubtless during that time have done much to stir up the Lewchewans, intimidate the authorieties, induce them to relax their non-intercourse regulations, and commence treating other nations more openly. We have made them receive pay for provisions, and gradually increased the amount of supplies until the ships began to get something nearer adequate to their wants; small purchases were daily made in the markets for the last week, and fewer of the spies tracked our steps, producing also less alarm among the women & children at our presence. The Chinese sent over from Shanghai to Dr. Bettleheim seems to be a man who will teach these rulers some new ideas on civil polity and foreign intercourse, and will less alarm their fears than a foreigner. He made his way into the palace last week, where he saw the prince & was civilly received by Máu, one of the treasurers. At a visit to the Mayor's, he was also respectfully treated. In breaking up the system of things so long upheld in this island, time & kindness, firmness and justice, united, & allowed their fair action, will soon have their due effect; we have begun, I think, in this manner, and I hope will not deviate from it, tho' I have great fears on the subject.

Aug. 2nd

This evening, to the gratification of every one, we met the "Vandalia" on her way to Napa, and obtained letters from her, among which I was happy to find one for me, informing me that all at Macao & Canton were in good health a week ago. It is something of an event for three U.S. men of war to meet in these unfrequented seas.

Monday, Aug. 8th

Last evening, the squadron anchored in Hongkong harbor, & to the regret of all, heard that the Powhatan had sailed for Lewchew Saturday morning; she would have been intercepted if the Mississippi had gone on ahead of us, as she might easily have done.

I find that friends are in general well. I mean to start this evening in a fast boat for Macao, having missed the steamer this morning. Thus ends the first act in the Expedition to Japan.

1854

Wednesay, Jan. 11th

Since I left the Mississippi & Susquehanna at Hongkong, I have been engaged in carrying on the Fan Wan[23], which yesterday reached the 400th page. In October, whileat Macao, either thro' exposure to the sun, or the effects of a cold and malaria, I was taken sick with a low nervous fever, which reduced me very much; it was the first sickness I have had since childhood, & I bless God for recovered health at this day, so that I can leave in health to rejoin the Expedition at Hongkong. I depart from my home in full confidence of my being where duty calls me, & leave my family under the care and governance of our heavenly Father, who has hitherto watched over us all. Mr. Bonney has, unwillingly, taken care of the office again until I return, after which, if I am permitted to do so, he will leave for U.S. I have secured the assistance of Lo[24], a teacher of good attainments & no opium smoker, so that I hope to do more study than I did before.

Tuesday, Jan. 17th

I am on board the Susquehanna on Friday evening, having learned that the squadron sails early in the morning. The officers all anxiously hoped that the mail [~~will~~] would come before the ships leave, but the Commodore [~~will~~] would not wait for it; [~~and~~] happily it arrived about ten o'clock in the evening. I saw the Bishop on Friday, & he wishes to hasten on Betteleheim's labors as a translator of the SS. [=Scriptures] so that the Bible Society can have somewhat to print. Mr. & Mrs. Morton expect to sail on Tuesday for Lewchew in a ship bound to California.

While in Hongkong I remained at Mr. Johnson's[25], where also the officers have often frequented & been pleasantly entertained.

All being ready, the ships weighed anchor about 9 o'clock on Saturday, and steamed out of Hongkong harbor, saluting Admiral Pellow's ship as the flag-ship passed her in return for her salute of 13 guns; the Powhatan took the Lexington store-ship in tow, & the Mississippi took the Southampton, and all moved out nearly simultaneously thro' the Ly-u moon passage.

Today we have passed the southern end of Formosa, progressing rapidly on our course; the sea is smooth, and a fair view has been obtained of the shore, distant about 2½ miles, which offers few signs of inhabitants, some cultivated & stubble patches, a house or two, and roads leading inland. Many places might be reckoned as villages from the smoke which issued from them, but they were far off & could not be distinguished. Most of the shore was covered with low woods, & large areas appeared as if untouched by man. The soil was generally good enough to produce grass or trees, & no bleak, barren patches speckled the hillside as about Hongkong. The hills rose gradually to the mountain ridge, one peak of which was estimated to be 25 miles off & over 3,000 ft. high, and doubtless constituted a portion of the chain which forms the backbone of the island. This portion of Formosa has been lately made infamous by the capture of the Larpent's crew, after she was wrecked, most of whom were hereabouts driven ashore and murdered by savage natives, a few having obtained safety among Chinese villages, & finally escaped to the "Antelope" as she passed thro' this strait in their sight. Such miscreants as dwell at this end of Formosa should be severely dealt with, and perhaps the desolate aspect of the apparently fertile coast may be owing to their driving away peaceable settlers, and still afraid of living within reach themselves. Some blackfish and two black terns were seen as we passed the straits.

Jan. 22nd, 1854, Sunday, Lewchew

We reached the harbor of Napa last evening at sunset, finding the Macedonian, Vandalia, & Supply at anchor here, and their officers pleased to see us. Mr. Betteleheim also was soon aboard, and reported that the authorities seem to have made up their mind to endure in our visits & remaining here what they can not care or prevent. Today has been a complete turmoil in the squadron from the orders which have come out from Washington to put one of the steamers at the service of Mr. McLane and take him about. The commodore moves himself and suite into the Powhatan, which necessitates some other changes and a good deal of work. Service was held today, and then the orders were made known, which of course set everybody a talking, and utterly destroyed all seriousness. Truly, the desecration of the Sabbath in a man-of-war is as great as in a pagan country, where it is not known, if we except the general cessation of work; but as to keeping it holy, I fear the thing is impossible if one wishes to do so, amid such a melée [French=medley]; no privacy can be obtained, voices are heard talking, however close one's door is shut, and the great majority have no possible way of secluding themselves from their shipmates if they wished to do so ever so much. Added to the bustle on board, a deputation came from the Mayor of Napa to salute the Commodore, and the their members wearied out nearly an hour in the captain's room, saying little, and making him (Capt. B.) nervous.

From them we learned that the old Regent is still living at Shui, rather infirm; that the new one & the Mayor are the same as when we left in August, and that junks begin to arrive from Satzuma in March. Towards evening, a present of a bullock, two goats, 2 hogs, 15 chickens, eggs, turnips and potatoes, came off from the Mayor to the commodore. The manner in which the Lewchewans tie up eggs in straw, by plaiting them lengthwise inside of alternating strands, is very pretty and safe, and prevents them breaking with usual care. These things were received, and their bearers at last went home.

Jan. 23d. Napa

This morning, I was early aroused by the noise of hammers and chissels [*sic*] and the voices of workmen, who, beginning to pull up the fastenings of the house and take town its partitions, gave me no peace. All this was preparatory to moving on board the Powhatan, where the Commodore & his suite are to remain until the cruise is over, as the Susquehanna is to return to Hongkong to receive Mr. McLane. We are all much inconvenienced by the change, and the artists more than others, as they are sent on shore to do the best they can at Fumai in the house hired there, cheerless and dirty as it is. I am meanwhile placed in the cabin.

About 10 o'clock I went ashore to return the Major's deputy's visit of yesterday, accompanied by Lieut. Brown as the commodore's deputy. We went directly to the Kung-kwan of Napa, instead of going to Dr Bettelheim's, and there waited two hours for Idzirazishi and the Mayor to come. Meanwhile, a pleasant man, whose ancestors came from Fuhchau about 120 years ago[26], made himself agreeable to us. At last, the officials came, & we gave him the Commodore's salutations, and told them our message, that he intended to take a trip into the country in a week to be absent three days or so, [~~and~~] requesting them to prepare a cortege of coolies, chair-bearers and guides, with 8 or 10 horses to ride on and carry baggage. These intimations did not at all please them, and various obstacles were interposed, such as the distance and a separate jurisdiction of the northern part of the island, over which the Mayor had no control. He was then requested to inform the proper authorities of the proposed visit, & furthermore to tell the Regent that Perry intended to see him & pay him a visit while in port. The interpreter hoped that the Commodore would pay this proposed visit [~~to~~] at the kung-kwan and not at Shui. This I said was against all usage, and could not be allowed. So we came away.

In the evening I went with Dr. Wilson of the Supply to see a neat little garden, made with coral in fancy garden style, in terraces and pools, with dwarfed pines and other plants, miniature houses and pavilions, all in the neatest style on an area of about 20 ft. square. Some gold fish and other kinds were [~~also~~] swimming in the lower pool. The inmates of the house were very affable, but we could not communicate much with them. I am told there are many such fancy imitations in Shui, all perhaps taken from larger Chinese originals. During the rest of the afternoon, we saw perhaps a dozen people transplanting rice, whh is here allowed to grow much higher before put into its new bed than in China. Dr. Marrow killed a kingfisher this afternoon of a steel blue & bluish green plumage, different in several points from those common near Macao. Also a plain gray-brown crane, which is common on the shore; it is two feet high, and has yellow irides, and is speckled black on the yellow legs; named ko-sáji, & the kingfisher is kánzúi.

Tuesday, Jan. 24th

This morning went again to the Mayor's at Napa, Capt. Pope & Lieut. Brown being deputies of the commodore's. The time of waiting was spent at Dr. Bettleheim's, whom I found most anxious to get away to China as soon as Mr. Morton comes. The message to the old Mayor was to ascertain the price of building the coal shed and cost of materials, so that the bill may be settled, and right of ownership established by the receipt of the authorities for it. Also to have him see that the horses, & coolies needed for the excursion be in readiness at Túmai. My teacher [Lo: 羅森] is greatly amused at these people, their beggarly equipages and aspect, the way in which they go about half dressed, and their unwillingness to sell provisions. One man told him, "What use can we make of your money? If you'll give us a piece of port we'll give your potatoes, for then we shall have somewhat to eat, but we can't eat cash." Thus the avaricious officials appropriate all the profits of the purchases of provisions for the fleet.

After leaving the Mayor's, we met by Idzirazichi, who said the Regent & Treasurer were in waiting at Tumai, unable to get off to see the Commander. I went alone to see him, and received the cards with the spirits & cakes he was to have taken off; further effort was made to deprecate the proposed visit to Shui, but I told them it was out of the question. Lo's idea that it would do these officials good to bamboo them to teach them manners, is not far from the truth. However, God's purposes may not yet be ready thus to deal with them; but their nonsense & prevarication are very provoking, while it is really, too, about the only weapon they possess. Active efforts to oppose us they cannot bring to bear, & passive resistance is their only alternative.

I dined today at Dr. Betteleheim's with Marrow, and passed the afternoon there. Mr. Crosby, the 3rd assist. engineer of this ship, was buried today at Túmai, where now rest some six bodies from the fleet, over all of whom, except him, the Lewchewans have built solid stone tombs & plastered them nicely, without any demand for payment.

Jan. 28th Napa

The three last days have been so stormy, & the swell & surf have rolled in so high, that few or no boats have left the ships, and very little work has been done. The Commodore gave a dinner to Capts. Boyle & Glasson of the store ships yesterday.

This morning as usual I went to see the effete old mayor of Napa, to urge him to do what I suppose he finds difficult enough, viz. to get our regrets fulfilled. The means of defense this people possess lie chiefly in their weakness, and in constantly saying that they have not this, & can't do that, & to weary us out by delaying and excuses. The Commodore wished today to get coins in exchange for Amn. coins we showed the Mayor, and straightway the querulous old man began to say there were no coins in the country, that the Japanese never brought any coins to Lewchew, where no one used them, & ended by declaring that as there were none, so none could be got. His assistants took an order, however, for a large chawchow box, 10 lacqd. tumblers & a punch-bowl of lacker, to be done when we return from Japan, which they made no objection to doing. It is exceedingly provoking to hear the lies and nonsensical excuses made by these officials, when all that is necessary is for them to let their people do as they please, tell all that we ask for, and keep themselves away. Another thing wanted today was the bill for erecting the coal-houses, and this too they boggled at, as if it was some new thing; when they learn to take our words just as we say them, there will be a great advance on present intercourse.

In the afternoon a large lot of presents were sent ashore —a box of drills, a dozen of champagne & cherry cordial, a box of 35$^{lbs.}$ sperm candles, & a box of Oulong for the regent, together with a small chest of tea for the first & 2nd treasurers; all of which valuable articles were delivered in exchange for the saki & gingerbread handed in by the Regent, & taken by the interpreter. I also told him not to fail in getting the coins, as we were determined to have them; indeed, I have an idea that a good deal of the hindrance we find is owing to this Idzirazishi, who may be compelled to this course by his superiors. The jaunt into the country is now delayed a few days.

[Sunday], Jan. 29th Napa

Altho' today is Sunday, there is little cessation from work or business, and if God adds his blessing and enables us to carry out the design of the Expedition, it will not be because or in answer to our prayers or regard for him, but because we are used, as Nebuchaduezzar [Bible: d. 562 B.C.], the ax[e]-helve, was, to carry out what falls in with his plans. In fact, no regard seems to be paid here to whatever scruples a man may have about doing work on the Sabbath. Mr. Brown went ashore to see the officials about wood, ballast, boats, & coins, all of them objects of minor importance, and easily deferable to another day. Dr. Smith was also ordered to go ashore to see about a man lying in the hospital with a broken thigh, which service was really no more called for than if he had been sent to see the condition of the boats lying on the beach. In truth, God's day, and by consequence others of his laws, are made to give in to the will of one man, or else the subordinate subjects himself to the penalty of disobedience or mutiny, of which every officer at least is very jealous of incurring even a suspicion of.

Jan. 30th

On going ashore with Capt. Adams & Lt. Brown to meet officers at the Mayor's hall, we saw many signs of the newyear like those known in China, among which the renewal of the papers with inscriptions on the door-posts, the numbers of well dressed people & children, & the clean streets, were the most conspicuous. The markets were generally open, however, and one or two mechanics were at work. The streets were not thronged as much as usual. At the kung-kwan, we met the Treasurer who had been waiting for us [~~a~~] [~~while~~], & had a session of two hours, during which we obtained a receipt for the erection of the coal-shed and for the rent of the hospital for six months to March 1st at $40 per annum. The cost of the coal shed was placed at $90. While other matters were talked about, Rev. Mr. Jones came in to engage bearers to take him & a party to the north of the island tomorrow, to investigate Lieut. Whiting's report concerning a coal and iron mine in the region of Port Melville. The facility of having higher officers in concluding matters was here seen, for our demand for boats to take off ballast, coolies to carry this exploring party, and to take pay for the coal-shed, were more fairly complied with. I told the Treasurer that the Lewchewans were as difficult to take money as a sick child was to take medicine. Our request to exchange coins was waived as before by a firm denial that any were procurable, altho' I adduced the proof of Japanese coins having been got of Lewchewans in Fuhchau; perhaps this demand trenches on their desire to disavow all knowledge & presence of Japanese.

In the evening, the interpreter came to Fúmai, & received the $110 for rent, &c., which was settled after a long discussion. In signing their names in Chinese, he & his fellow, added a rubric very much like the Spaniards, formed of two or three characters; it seems to answer instead of a seal.

Jan. 31st Napa

I was employed on board all day preparing a document to take to the Regent himself, in which the Commodore takes a firmer stand, and tells the officers of this petty island that he can no longer submit to their subterfuges and nonsense. One cause of this move is, that he sent off his steward this morning to get some fish from the boats out on the reef, and they fled; and on going ashore, he fared no better, as there were none in the market; so the Commodore, instead of fish for his breakfast, got nothing but a blue-slate crane, which his messenger had picked up somewhere. Furthermore, the demand he has made for coins has been met with a firm denial, that no such things are known or brought except cash; while he learns today through a native who has thrown himself on our kindness, and paddled off to the ship Sunday night, desiring to go off with us, that many Japanese coins are brought here, tho' they are not in a circulation. Again, Liuet. Whiting has brought a specimen of powder he procured at a mill he came across in his survey, while all kinds of arms & powder have been often asserted to be unknown in the island. Taking all these things together, the Commodore is going to talk "strong" to them, and see what effect it will have, especially as he is soon to leave for Yedo, & all that we do here is reported there, and may influence our reception there.

Feb. 1st Napa

Early this morning, the marines were sent ashore under Capt. Slack's order to drill, and Lieut. Brown, Mr. Perry & I were off by 7¼ A.M. to take the Commodore's letter up to Shui and give it to the Regent himself. We met the marines near the bridge, and joined by Mr. Eldridge went up to Shui with them. As we neared the capital, the music and arms of the men attracted attention, and the people came running out to see the show; but it was when we saw the Regent & Treasurer coming out in a great hurry to see what the matter was, that the extreme confusion this sudden visit had thrown them all into was best exhibited. They conducted us four into the hall, and began to make preparations to make the empty chamber fit to receive us, mixed up with questions to us, orders to the servants, & a half unsatisfied, terrified, air, which showed how scared they were. I gave them the paper, told them the Commodore was coming on Friday to the capital to pay his respects to the Prince, and wished them to have horses & chairs & bearers in attendance at Túmai. They made apologies, & hoped that the Commodore would receive his visit at Napa, for the prince was young & his mother was sick, &c., but we got up to leave, declining their refreshments and reiterating the orders we had received. They made efforts to have us stay, and had not fully recovered from their alarm when we came away, but as there was nothing to be said more, it was thought best to decline. The marines had gone on up to the palace gates, where a large crowd was gathered to see them; and we told the officials we had nothing to do with their movements, that they had come ashore for exercise & marched up to Shui to entertain the people of the capital with a newyear's show. On our return, we had got nearly half-way back, before we heard the music striking up, and this mingled with the pleasant breese soughing thro' the pines, and at intervals the sheen of the guns and uniforms as the company came in our sight, rendered it a very pleasant & pretty show. I expect the effect on the officials will be salutary in a reasonable degree, and make the people used to us.

In the afternoon, I strolled thro' the streets with the teacher, and found our way into a number of places he had not seen before, one of which was the graveyard near Dr. Bettleheim's. In this place, most of the tomb-stones were placed on pedestals, each monument being in the midst of a square inclosure made by a low stone wall; the stone was soft, fine red-sandstone or a whitish rock like tufa. Some of the epitaphs were dated in Kienlung's or Kiáking's reign, but all the recent ones were dated in the reign of the siogouns Kayei, Tínpò, or others, from which one might infer that Japanese were buried here, or that a stricter oversight was taken of the acknowledgement made of the Chinese by the Lewchewans. Most of them commenced with 歸元 or 歸真 or 歸空 , i.e. "returned to his original", or to "certainty", or to "nothing", "emptiness, ["] annihilation", I could get no one to tell me about them, but the epitaphs indicated official rank. The oldest grave was not over a century. A few had hirakana writing on the side of the intaglio-cartouch containing the Chinese inscription; & one or two others were wrought into a square pillar placed on its end, & surmounted with a roof, all of stone. The common style of inscription is here given:

乾隆二十一年
歸真惟安澈心居士靈位
四月十四日

The seat (or throne) of the spirit of the retired scholar Wei-ngán, styled also Chech-sin, who returned to certainty on the 14th of the 4th month in the 21st year of Kienlung.

The temple near by is a small building, and at this time was filled in its principal room [~~that~~] by a dozen or more men who seemed to have nothing to do better than to smoke & look at each other. The walls were hung with a variety of tablets such as are common in Budhistic establishments.

Feb. 2nd Napa

I was ashore this morning to see the Mayor respecting the visit to Shui, when I was told that the Regent & officers were about disembarking to go to the flag-ship and had to hurry off across the salt-pans to see them before they left. I was in time, & got into the boat with them, and we had chairs arranged for them on deck, as Perry declined to see them. They brought a reply to the paper taken yesterday, in which they promise to order the people not to run away, to supply all the provisions sent for, and to act with truthfulness, in which last they have promised more than they can perform, I think. They made many excuses for not being able to let the Commodore see the prince or dowager, and although they were willing to let him into the palace, it was inexpedient for him to see them; & it was not until this was agreed to, that they were made easy. There must be some reason for this difficulty which they do not like to let us know, perhaps because it verges too near to the Japanese rule. There was as much difficulty in this respect now as there was last summer, & perhaps it has been made more stringent upon them, since then, altho' from the description of Betteleheim's Chinese, who saw him last August, there is such a person, & I am told he often goes abroad. However, it was agreed to, & at this they left in better humor than they came up.

Feb. 3rd Napa

The morning was so threatening, that it was not till nearly seven o'clock that the Commodore concluded to go to Shui. The Marines were sent ashore immediately, and Perry left at 8½ o'clock, with the promise of a fine day. Eight stout fellows were on hand to carry him in his sedan, and we started at 9½ o'clock; the number of officers was less than at the visit last summer, and the absence of the field pieces made it a less imposing escort. The Regent and his associates received us in the palace, where the[y] had laid out a number of tables in the same room we were received last summer. Formal professions and salutations were exchanged, and the Commodore brought out his coins for exchanging with them, the Regent evidently unwilling to receive them; there were 9 sorts, valued at $49¼ in all, and they said they would do all they could to get their weight in Japanese coins, but declared to the last their non-possession of such, & the difficulty of doing as we desired. After a few other compliments, we rose & went off to the Regent's, no Prince being brought out for us to see.

At his residence, we found the tables all laid out for a dinner, & the various dishes brought on, to the number of 12 or 14, proved to be far more palatable than any we had previously tasted. They were all cooked as stews, and there was great similarity in the gravy, but not so much in the viands. The whole went off in good humor, and we left on our return at 11 o'clock, the Regent accompanying us to the outer door. The walk back was a delightful one, the fresh air and charming scenery exhilarating us all; in fact, no one can get tired of the views on the way up to Shui, and the industry of the village is nowhere better seen. After we reached the ship, the presents made to every guest were assorted, fans, tobacco pouches & paper to each, cloth, tobacco-leaf &c. to the officials.

In the evening, I took a stroll with Dr. Smith, visiting the markets, and finding the old women well disposed to sell, & one would be disposed to buy if they had anything worth having.

Saturday, Feb. 4th

I was ashore at the Napa kung-kwán today to see after the Japanese coins, but none were to be had; they declare that they have none, & I begin to believe them. They soon produced a bill of charges for the supplies & labor furnished Lieut. Whiting in his survey at the north, amounting in all to $108, more than as much again as he had judged. So, if they will not let us have coins, they are learning to like our's, and to charge round prices for all their little island furnishes. In their reply to the communication taken to Shui on Wednesday, they say that Lewchew is a "little out of the way island off in a corner of the sea", and entreat the Commodore in the greatness of his kindness to have compassioned on them, & promise to do all he wishes; but his table is all subterfuge, and may be taken for nothing. However, as it was, I took advantage of it to get some supplies for the Engineer's mess & some sailors from the Mississippi, telling two of the sailors to go into the market & get some [~~hotes~~] for themselves, which they managed to do. The sailors have contrived to supply themselves with many things from the markets during the time we have been here, & have quite brought over the old women there to think they are good customers.

Sunday, Feb. 5th

I was unable to move to the Powhatan yesterday, and I had hardly gone aboard that ship, when I was sent for by the Commodore. I took a final order for the Regent to have the coins ready on his return from Japan, telling him that he would know whether his professions were real, by his getting them. I knew that Lieut. Brown had gone for them, to make a last trial, and was quite sure he would not succeed. On reaching the flag-ship, I saw a large number of presents with cards in return for those handed in on Friday, spread out on the deck; these were in return for the barrel of whiskey & flour, & a lot of garden tools, given them yesterday. They were all received, and the boats had left for the shore, when Lieut. Brown & Idirazishi came off, bringing back our coins, & a paper from the Regent stating his inability to obtain any in exchange. The Commodore declined to see them, but on hearing the paper read, ordered all the presents back into the boat, and gave them his own communication to take to the Regent with the coins he had given him at the palace. In doing this, I think Perry acted like a disappointed child, and was piqued at being unable to effect the exchange of coins he had set his heart on. He bids me tell them that he asks only for what is reasonable, and that the exchange of national coins is a sign of friendship; these islanders are known & allowed to have no mint of their own, but a breach of amity is made to depend on their furnishing the coins of another land, which they deny to have or be able to get. I think this matter was carried much to[o] far, & as I will tell no lie for Perry or any one else, I never told them he asked only for what is reasonable. I was much vexed at the rejection of these sundries, & hope the Regent would send the shovels & hoes, flour & whiskey, presented to him yesterday, back in like manner. He has doubtless exerted himself, & can do so still farther, & it was well to leave our coins in his hands; but that matter, as Idzirazishi said, had nothing to do with the presents sent. If the coins desired were Lewchewan, the case would be materially altered; as it is, I think Perry is in the wrong in pressing the exchange to such a degree.

Towards evening, the wind rose to such a height that we shall not leave in the morning; and thus the violation of the Sabbath in this needless, sinful manner has not advanced our progress one hour. But many ask, do you think the wind would not have blown just as hard if we had been at prayers and preaching all day? I can not say what would or would not have been, but I know that God can thwart any plan which involves infraction of his laws, & that the obedient way is always the safest and surest in the long run. Dr. Bettleheim preached on board the Mississippi, but there was no service in the other two ships.

Monday, Feb. 6th

A signal was made for me this morning early, and

on reaching the Susquehanna, I found Mr. Randal there, and the list of the presents brought off yesterday lying on the table. He said the articles themselves were then at the house in Túmai, and that the interpreter had been with him yesterday, apparently very desirous of getting off to the ship. On going ashore with him, I sent for Idzirazishi, who came after a long delay, & said he had nothing to do further in the matter, that the list had been brought there, & that the communication had been sent up to Shui, but no coins had yet been procured. However, we got pigs, poultry, & potatoes, which are now more valuable than coins, in my opinion, the purveyors having bestirred themselves on seeing their profits were departing, and brought down the largest lot of eatables to the beach that had before been seen in Lewchew. I pitied this interpreter, for I doubt not he is in an unpleasant dilemma, and would willingly sell all these presents to relieve himself from the difficulty of taking them back to Shui. He perhaps states things pretty much as he desires them to be, and a course of such conduct soon brings him into some troublesome explanations; tho' in the main I doubt not that he is honest.

The Commodore thinks that as he has once set out to get the coins, & believes that they are to be found in the country, it will not do to retreat from the attempt; and such a determination is the best way of dealing if we were sure the coins were to be had by them after a little pressure. The matter is now left until we come back, & I am inclined to think the authorities will try to get some if possible.

After dinner, half a dozen of us went to Shui, attracted by the pleasant afternoon to take a stroll. The country looks exceedingly pretty, freshened as it has been by the recent rains, and brought out in all its beauty by the high cultivation it is under. We got up to the highest part of the castle walls, and enjoyed the pleasant view in all directions by the light of the setting sun, which cast a cheerful glow over the charming landscape. Few prospects could delight the eye more, but how great an increase of interest would be given to it, if one could feel that these villages and towns, were the abode of a Christian people!

Tuesday, Feb. 7th

This morning all the steamers were under way betimes for the land of the Rising Sun, and we had hardly gone fifty miles before the "Saratoga" came in sight, she having been seen, it was supposed by a light, the night before, when a gun was fired from the flag-ship. After a short stoppage, and a visit to the Susquehanna by the [~~Commander~~] captain, she went on to Napa, leaving some of her cattle & sheep on the Mississippi.

By her I had a line from Shanghai, which the rebels still had in their possession, but trade was going on pretty much as ever, fighting on one side of the town & trafficking on the other.

During the latter part of today we came in sight of O-sima, the large island lying north of Lewchew, and appearing in its general features not unlike that island, low, wooded, and cultivated. Whether it supports as dense a population is doubtful.

Saturday, Feb. 11th [Off Yedo Bay]

We have been highly favored in our trip to Japan, having had smooth seas, and for a part of the way fair winds. The high land of Idzu and the islands off the Bay of Yedo came in sight this afternoon, and a patch of rocks too which are said not to have been seen on the former trip. The winds have become very piercing, though the thermometer indicates not much more than 40° Fah. As it was towards evening, the Commodore lay off the mouth, drifting about until the morning, altho' the moon gave sufficient light to see the land by. It is a bitter night in the rainy, driving north winds, and we ought to be thankful for protecting mercies.

Feb. 12th

The steamers were all pressing on towards the land, which was almost everywhere white with snow on the hills. As we neared it below Cape Idzu, we erelong descried two of our ships, and ran up a coast none of us were at first able to recognize; and judging that Oö-sima or Volcano I. was correctly laid down, supposed it be off Cape King on the eastern side of the Bay of Yedo. It was ascertained, however, after a time, that we were in the Bay of Simoda, where the British man-of-war Mariner anchored in 1850; and where too we soon learned that the Macedonian was ashore. Consequently she must first be got off, and this the Mississippi did, [~~and~~] dragging her into deep water, when all the ships lay for the night where they were, the Lexington coming up in the evening from sea. Some towns of considerable size were observed along this bay, but not many boats were seen, owing probably to the cold wind, deterring all coming out for mere curiosity. The news of our arrival was perhaps made known by some of the fires we saw lighted on the beach & hills; but more likely by couriers started for the capital.

Feb. 13th At anchor above Saru-sima

During the night the wind went down, and the bay became smooth as could be wished. Towards sunrise, one of the most glorious scenes [I] ever beheld was to be seen by those who were up, but I was not out till after sunrise. Mt. Fusi lay right before us, clothed with a pure mantle of snow, & all the high points of the landscape were of the same dazzling white, including the island of Oö-sima, from which the smoke now could plainly be seen rising, & settling in the lustrous cloud above the summit, thro' which the rays of the sun shed a peculiar brightness. The shores of the bay were destitute of snow, and the dun brown of some parts with the dull green of the pineries, added other contrasting shades to the snow, rendering the whole variegated and beautiful. As the sun rose to view, the tops of Fusi & other hills were touched with a roseate hue, which disappeared as it came further up, but the brilliancy of the whole compensated for this transitory charm. It was a magnificent sight in every respect. I could not help praising the still more glorious Being who has given us senses to enjoy these beauties, as well as made them to hold him in remembrance by the children of men. Truly his works praise him, whether we do or not; and we are the more guilty if knowing him we do not exalt his holy Name in view of these manifestations.

Feb. 13th American Anchorage

By noon the six ships, each steamer towing a ship, were off the town of Uraga, but the Commodore passed on until he came to the American Anchorage some miles above Saru sima or Monkey I., where we all anchored, the Southampton having been already here three days, and often visited by the officials, with whom Capt. Boyle managed to communicate. The coast was destitute of snow, but its bleak, dun [~~aspect~~] color gave it quite a different aspect, so that one might well doubt its identity.

In the afternoon, after having dodged here & there for an hour or two in pursuit of us, the Japanese officials came off to us, four in number, two military men whom we saw last July at Gori-hama, and the two interpreters. They came to the Powhatan after having visited the flag-ship, & were received in the deck-cabin by Capt. Adams. Their chief object was to inform us that a person of higher rank was coming aboard tomorrow to consult respecting an interview and the reception of the emperor's letter. They wished to know why we had anchored so far above Uraga, from which it was a long way for them to come, and desired us to go down off that town at least in one vessel, so that we should be more accessible; this move was declined on account of the more secure anchorage at present occupied, where there was no fear of winds. Their proposition was made evidently only to make it more convenient for them to get off to us, though a dislike of our going nearer the capital may have had its influence. In windy weather, it would be a dreary sail for them to come up in open boats from Uraga, tho' they can come a good way by land, & save the boating. They left in good spirits, and towards the last intimated that the answer to the President's letter would be a favourable one. The number of attendants was greater than Yezaimon brought with him last year, but all equally well bred, as those.

Feb. 14th American Anchorage

The official spoken of yesterday came about noon,

with two colleagues. The name of the first was Kurokawa Kahiyōye 黒川嘉與勝（クロカワカヒヨウエ）and as near as we could ascertain, he fill[ed] the post of prefect in the principality of Idzu, resident at U[ra]ga, a higher officer than Yezaimon, and from the imperia[l] coat of lily-leaf arms worked on his breast under the ov[er] tunic, perhaps connected with the supreme government [as] a deputy on its part in this important port. His coadjuters were called Yoshioka Motohei 吉岡元平（ヨシオカモトヘイ）[and] Hirayama Kenzirō 平山謙二郎（ヒラヤマケンジロウ）, whose official p[osi]tion we did not learn, but one of them was evidently no equal in counsel to the principal man. More attendan[ts] came than yesterday, one of whom was a lad, who main[tain]ed his post close by Kahiyōye amid all the confusion, [hold]ing his master's long sword bolt upright in his hand, d[ur]ing the long interview. It reminded one of the pages [of] the middle ages, whose duties comprised such services.

After accommodating the party with chairs as well as we could, & some of them with other conveniences too, the interview commenced with their making an apology for not coming sooner by reason of the distance from Uraga, and begged us to move at least one of the ships down opposite that town for convenience of their going to & fro. It presently was evident that these officials came to arrange about an interview on shore with an envoy from Yedo, who they observed had the answer from the emperor to deliver; and that as the President's letter was of so polite & pacific a character, the reply would likewise be favorable, and they hoped we should be able to arrange amicably for this interview. They then said that Hamakawa, a town in the bay of Simoda, near where the Macedonian grounded, was a very convenient and large place for it, but this town was decidedly rejected. At last, they intimated that as the town of Uraga had been selected by the gov$^{t.}$ at Yedo, they had no authority to change it; tho' they had come to settle the preliminaries of a meeting there. They urged that as we had made no objections to the place of meeting last season on a barren, uninhabited beach at Gori-hama, & had delivered the President's letter there, and that Uraga was a large town set apart for the proposed interview, where officials resided, and where it would be more convenient for them to prepare for it, all propriety was in their favor, & we ought to accede to them, & at least move one ship down off Uraga. We proposed some place between the ships & Yedo, to which they declined, alleging that there was none suitable; and said we would state all they had said to the Commodore, which Capt. Adams did by note. This gave a chance for a respite, & they all got up to see the engine and other parts of the ship, in which some took an interest in one thing, some in another. Our first visitor of last year, Saboroske, was here today, & took a minute admeasurement & plan of the big gun on the quarter-deck; he seemed to be a secretary, & had a convenient set of writing tools with him, which he used in his hand. Others also had these portable writing tools. A few of the visitors came into the ward-room, where cake & wine were given them, most of [~~which~~] the first they wrapped in nose-papers to carry home. A comparison of swords with our's was then made, and they seemed pleased that their's were the sharpest. Many objects of wonder to them were exhibited, but they repressed all exclamations of surprise, and talked little among themselves.

After a while, we were all again seated, and as Perry had refused to stir lower down, Capt. Adams got them to take the proposal ashore that an officer should be appointed with whom he would go ashore and select a suitable place, and they might return with the answer in two or three days. We suggested Kanagawa far up the Bay but that was negatived, and every period brought us around again to Uraga as the appointed & best place. If we had had one of Joshua's 24 hour sunshine days, they would doubtless tarried longer; and seemed at last dissatisfied with our refusal to go down the bay, and take up with their place for meeting. They talked a good deal among themselves, but never confusedly, waiting each on the other, the two principal ones doing most of the confabulation.

Among other things they said they hoped no surveying parties would go out while negotiations were going on, but this desire could only be referred to the Commodore. We had a good deal of sport in exchanging cards & autographs, for which they seemed to have always a strong desire, according to all travelers; their cards are always in running Chinese characters, if these are to be taken as samples, from 2 to 3 inches long, and 1 to $1\frac{1}{2}$ inch wide. A variety of articles were placed in their capacious bosoms, into which they found their way by putting their hand back in their sleeves. Some of the party had eight or nine garments on, one over the other, and all were clad warmly, and all bareheaded. In the course of conversation, the interpreter said that they understand that I was not coming back this year, but I have no idea how such an [~~idea~~] impression was received by them. On the whole the interview passed off pleasantly; and [~~the~~] [~~whole~~] all [~~of~~] our visitors were apparently gratified at what they saw.

Feb. 15th Bay of Yedo

Preparations are making on board for receiving the Commodore on board this ship, but he is just now too unwell to move about much; it is rather inconvenient for him to be in another ship while negotiations proceed in this one. The weather is pleasant now, cold enough to make it agreeable walking on deck for a long time, and yet not too cold for writing or reading.

The younger interpreter came today about two o'clock with a party of friends, most of whom had not been aboard before, & whose object was chiefly to see the ship. Among them was a third interpreter from Nagasaki, who spoke considerable English with a good accent, tho' he did not talk much until we began to go over the ship, when his curiosity was so excited that he had questions to ask as well as much to see. The official part of the vist today was to inform Capt. Adams that in case he wished anything, as wood or water, or to call for other officers, if he would send a boat inshore, persons would meet him & convey the message. No answer was returned about meeting a deputy to consult with him relating to the place of interview, and the visit was rather uninteresting. The forward deck was well crowded while the Japanese were looking at the guns, and another comparison of swords was gone thro' with: they hold our's in small esteem from their being so dull, regarding the metal as inferior.

Towards evening we remarked a large number of boats anchored inshore, at intervals, & a few outside, the whole looking like an attempt at placing guard-boats around us. If they persist in this, there will be cause of trouble found erelong, I fear, for not to do something will render the boats ridiculous in their own eyes. The number of boats seen in the bay during the day fully equals the number seen last summer; but we are now out of the way of the ferry which plies across from Uraga, and only a few come around to see the ship. The gulls hover around the fleet in numbers, attracted by the offal. On shore we can see the people cutting grass and faggots of bushwood among the pines, bringing them off in all probability for firewood. No snow lies on the shores any where in sight, but the mountains in the distance NW. are snowcapped, & almost rival Mt. Fusi in elevation.

Feb. 16th Bay of Yedo

About 2 o'clock Tatsunoske and a party of gentlemen came on board, none of whom were before in the ship, to my knowledge. The leading officer was a pleasant but forward man, and had almost nothing to communicate, the principal object of their visit being to see the ship. They told us that the high officer had not yet come from Yedo, but desired us in strong terms to move back to Uraga, where communication could more easily be carried on. They said they were sent to beg the Commodore to take care of his health & to inquire after it. In due time, the chief man brought out from his bosom a parcel of navy buttons, which had been given yesterday to the boatmen alongside, & had been taken from them, & now returned in this ridiculous manner. I asked them if they deemed us to be children that they trifled so with us, and told them such was not the conduct of men & friends towards each other. They told me today that the siogoun had died in the 10th month of last year, but that the nengo nengo [年号] of Kayei was still continued, & this was its 7th year, altho' a new incumbent had the seat. Such a mode of reckoning must throw history & chronology into some confusion; & it shows too the duplicity of the people, for no other adequate cause for such a step can be assigned than to deceive by confirming the impression that the same monarch still reigns. A day or two ago, one of our visitors told me that the mikado had resigned, but I did not ask him then whether the siogoun was dead, supposing from the nengo being the same that he had not died, as we heard reported. One of our visitors, today, was 71 years old, & I observed that the Japanese have the same habit of showing their fingers to indicate small numbers which the Chinese have. Most of the time till after 4 o'clock was taken up in walking about the ship, in visiting the wardroom, where Dr. Maxwell tried unsuccessfully to electrize them, and others showed them pictures, swords, pistols, & other things to entertain them, and in examining the machinery. Their numerous inquiries to see the engine indicate the interest it has excited, and I told them that we reached Yedo, the emperor & his councillors must come off to see it also; the look of doubt & surprise was all of their answer which came outside. The manners of the chief man, whose name I've forgotten, were so pert, and he was so disobliging, & acted so silly in relation to the buttons, that the officers in the ship are not inclined to show them any more civilities when they come aboard, until they exhibit some desire or intention to reciprocate. This was particularly offensive in this man, when he refused to show a pretty bag hanging at his girdle, containing some kind of medicine.

There was a grand review and inspection of boats today, which afforded all some entertainment; perhaps 400 men were prepared for action, and the whole fleet made a pretty show.

A surveying party went out today, a few of whom stepped on shore, and others had intercourse with native boats. The conduct of the Japanese will be tested, as this survey of the harbor proceeds.

Feb. 17th

Today was a rainy, chilly day, & no visitors came off, nor were many boats seen in the Bay. Whales frequently appear in these waters, probably cows which come in here to calve; some of them have been seen 40 feet long. Gulls of several colors constantly play around the ship, attracted by offal. On shore, the young wheat or some other green grain, begins to revive the summer garb.

Feb. 18th Bay of Yedo

A small party, of whom Yaboroske was the chief man, came on board about one o'clock today, bringing among other news the information that one of the chief councillors & his two coadjutors had arrived at Uraga, and wished the Commdore to go there and receive him and the reply to the President's letter. The same reasons were adduced, & the same objections brought against this step, as had been repeatedly gone over with; on this occasion it ended by giving them in writing the refusal of the Commodore to go down there, but he would send a ship and bring the commissioner up to this anchorage. The name given to the commissioner today differs from that handed in on Wednesday, which was Lin 林 [林 大 學]; now it is Hirayama Kenzirō 平山謙二郎（ヒラヤマケンジロウ）and his title is less exalted, being styled Revisor of Documents and general Counselor and Director of Affairs of the Frontier of Japan, 日本國監察参謀兼掌邊陲事カンサツ サンボウ ケン ショウ ヘンマイジ his colleagues have not titles, at least none given to us; their names are Yamamoto Bonnoske 山本文之助（ヤマモトブンノスケ）and Mayeda Yōtarō 前田右太郎（マ

エダユウタロウ）. One might infer from the title of the Commissioner that it was given him for this occasion.

Yaboroske brought a box of confectionary today, which consisted of a few varieties of [~~jellied~~] [~~syrups~~] [~~and~~] candied jams of fruits. His activity of mind is remarkable, and he improves on acquaintance; today he took the measurement of the ship with a fish-line, & has previously taken dimensions of the guns on the deck & their appurtenances. He seems to be a secretary of Yezaimon, but does little else at our interviews than take notes. When the party left today after taking the written paper, Tatsunoske was told that the morrow was one Sabbath when we did no business. From their conduct today, there is doubtless a decided determination to get us back to Uraga, and we shall perhaps have to give in, and go down there.

The Commodore moved aboard the Powhattan today, but was too ill to do anything, and suffers a great deal of pain, the results of his cold caught on the passage up the Bay. The Southampton went up the Bay some 5 or 6 miles yesterday evening, to assist the surveying parties.

[Sunday], Feb. 19th

I attended service in the Mississippi this morning, and heard Old Hundred sung nearly by all the ship's company. Notwithstanding our desire for quiet, Kahiyōye came again about two o'clock with a party, bringing a present of radishes, greens, eggs, chickens, oranges, confectionary, and onions, altogether amounting to 3,000 articles & over. They were given some tea & biscuit in return, which they accepted willingly. After a good deal of circumlocution, drinking, walking about, counting the articles brought, looking at pistols & pictures, and doing other unimportant things, the rest of the party left the room without apparent cause, Tatsunoske alone remaining, who drew his chair up, and told us confidentially that it was the express command of the Siogoun to the commissioner that the interview should be at Uraga, & all those interested in the matter on their part hoped no impracticable obstacles would be interposed, to this plan on our part; for as the emperor at Yedo was willing to grant all we asked and permit a trade, this opposition would only impede what otherwise was likely to go on amicably. They understood our reasons for not wishing to move such ships into dangerous places, & would state them to him, and also our proposition that he himself should apply for further instructions to Yedo, if it were possible to have the place of meeting elsewhere. This colloquy was ended by our request that what he had now told us might be given in writing tomorrow or next day, addressing Perry directly, so that he might have a reliable document. They soon after all departed, leaving us under the impression that we shall obtain great part of what we ask for, & this large cumshaw [Pidgin English from *chin*= tip] of provisions increases this view.

During the afternoon, one of the Japanese complained of colic, and Saboroske took out a small box of tutenagen having 3 compartments in which were gilt pills, salts, & other medicines, the neatest homæopathic arrangement you ever saw. Taking another pill, he mixed it in water, gave it to the patient, who soon felt relieved; it was perhaps a preparation of opium. The skill of the man in preparing the dose showed that he was no novice at it.

Feb. 20th Bay of Yedo

The surveying boats have had considerable friendly

intercourse with the people along the beach and in boats, today and on Saturday, & erelong there seems likely to spring up a pleasant understanding. The people are evidently willing to cultivate kind feelings with their visitors.

Kaheyōye and his friends came again today, bringing a letter from Commissioner Lin and a number of his colleagues, in which they desired the Commodore to go to Uraga; and in reply he proposed to send Capt. Adams down in one of the ships, and bring them up if they wished to come. He stated his intention, also, in his reply, to take the ships up the Bay to safer moorings, and added that as he was sent to Yedo by his government, to Yedo he expected to go, where also he could show the presents sent out, and exhibit their mode of manipulation. They agreed to Capt. Adams going down. Kaheyōye also brought a cut shell as a private present to Perry, who returned a lithograph of a steamer, that seemed to please the official much. A[sic] 100 oysters in shell were also brought for Capt. Adams. The day passed off pleasantly, and they seemed gratified at the prospect of an amicable settlement & the opening of intercourse. Truly, we may say that God has gone before, & prepared our way among this people, and I hope it is to be for their lasting benefit too. If a place of meeting is appointed further up the Bay, we may hope to reach Yedo, the goal of our expectations.

Feb. 21st Bay of Yedo

A deputation of some low ranked officials came off about $7\frac{3}{4}$ AM. to see if Capt. Adams was going to Uraga, and to accompany him there, but I did not learn whether they intended to take him with them. They wished me again & again to go with them, taking me by the sleeve, and wishing to ascertain the reason for not making one of the party. The Vandalion got down near Saru sima, where she anchored in the afternoon, a violent storm of rain preventing further progress; so that the commissioners are likely to be kept waiting longer than they perhaps wish. The desire of these officials to get an interview at Uraga indicates the mind of the court, I think, not to do much to assist us to reach Yedo, knowing perhaps that they cannot make us as "respectfully submissive" as their Dutch visitors, and fear that they will thereby lose caste among the people.

The people seems to have no such apprehensions, and an intercourse has commenced among them from the boats sent to survey, which is plainly a voluntary exhibition of their goodwill and laudable curiosity towards "far-traveled strangers". The Camellias forty feet high, chestnut trees, a species of Laurus, pines, cedars, & other new plants to us, all possess unusual interest on entering a land so long shut out. In fact the gradual entrance into so peculiar a land in the way we have come, one thing opening after another, is not the least of the charms of the Expedition.

Feb. 22nd Bay of Yedo

I thought that we should be unvisited today, but a large company of gentlemen came on board about $9^1/_2$ o'clock, a part only of whom had been here before, to hear the salute fired by the ships on Washington's birthday. They rambled about as they pleased, and all seemed disposed to be entertained. One who had often been on board showed me a book of 20 leaves, giving an account of cannons, guns, revolvers, swords, and other arms, illustrated with neat and accurate drawings of each, diagrams of their various portions, so that a clear idea could be obtained of each implement. It was printed at Yedo last October, and I imagine that much of the information in it is a digest of what was seen aboard the Susquehanna last summer, tho' the author must have had some European work on gunnery to copy his drawings from. It was neatly printed, and the owner declined to let me have it on any account. He was carefully examining the guns while going thro' the ship. I endeavored to make the principle of the telegraph, which was set up to day, & in good operation, intelligible to one or two of our visitors, and made them comprehend that ideas could be conveyed along the wires by means of the machinery now exhibited, but how it was done was the mystery, which their partial knowledge & my inaptitude on such a topic, could not reach. However what was understood is likely to arouse attention.

The party which went ashore found kind treatment, and people of all sorts were curious to see the strangers. They went into a large village, where the women were not behind the men in curiosity. Some laborers in a quarry were tattooed or marked on the right shoulder, which they pointed out as if it was a distinctive mark. The island we call Webster's I. or Natsu sima, is uninhabited, but affords a pretty ramble. The village of Kanezaō lies inland west of it, & perhaps is the one visited today. The general condition of these villagers is not so comfortable, our officers think, as those Chinese who live about Canton. Houses are neatly thatched, mostly of wood or mud.

[~~Chinese short sentences with English translations~~]

[Inserted on the page 158 of the MS journal, Chinese short sentences and the English translations for them; in this volume there is given only the transcription text of the SWW's English translations, as below.]

The Great Mirror of Knowledge says, Wise men & fools are embarked in the same boat; whether they are prosperous or afflicted, they both are rowing over the deep lake; if the brilliant sails lightly hang to catch the autumn gales, then away they straight enter the lustrous clouds, & become partakers of heaven's knowledge.

To enter the abodes of the perfect, & to sympathize fully with the men of world belongs to Budha. It is only thro' one vehicle we can enter Hades; there's nought like Budha, nothing at all.

He whose prescience detects knowledge says, "As the floating grass is blown by the gentle breeze, or the glancing autumnal ripples disappear when the sun goes down, or the ship adrift returns home to her old store, so is life: it is a smoke, a morning tide.

We of the human race, having hearts, minds & understandings, when we reach the realms of Budha, how great is our advantage! Buddha himself earnestly desires to hear the name (of this person), & wishes that he may enter life.

The Canon of Budha says, "All who reach that land will become so that they cannot be made to transmigrate or change for the worse. He who has now left humanity is now perfected by Budha's name, as the withered moss is by the dew".

Couplets taken at Newyears from the doorposts in Lewchewan houses at Napa.

[Inserted on pages 159—161, Chinese couplets and the English translations for them; in this volume there is given only the transcription text of the SWW's English translations, as below.]

As the wind & light go their circuits thro' the world,
So does the gladsome spring from heaven to us come down.

The peach in fairy land ripens in thirty centuries,
May this seaside house have ninety autumns added to it.

We joyfully hope the flowery year will be a flourishing time,
For the bloom of spring we think has come to this humble door.

When the royal court is just, letters greatly flourish,
Spring comes alike to all, for heavenly powers have no partiality.

Once I hope his majesty may live for ever,
And then I wish his ministers may have a thousand years.

May the star of luck shine here for aye.

吉
照　拱
星

(Over almost every door of the respectable class, a board is hung with four characters arranged in this form.)

Without, holiday robes proclaim the happy age,
Within, the sounding pipe praises the royal merit.

Good luck daily enters the abode of peace,
And honors evermore come to the gate of prosperity.

Our joyful wish is to have the star of luck come to you,
As we see the gladsome spring is now before the door.

Spring descends from heaven,
Favor comes from his majesty's side.

Gay clouds (i.e. all joy) meet the rising glorious son [*sic* / sun],
Ten thousand joys greet the come-in spring.

Royal favor is like heaven for extent & greatness,
Like the lustrous, glorious sun, so is an upright rule.

The genial breath (of spring) envelopes all things,
And its balmy gales impart great pleasure.

With fair winds come great luck;
May you have peace & joy where'er you go.

Heaven confers peaceful, contented joys,
With man are found times of riches & honor.

May all joys wind around your blessed abode,
And a thousand lucks collect upon this gate.

Spring hies us off to the Great Yu's domain,
And then men feel as in the days of Ancient Yán.

May your felicity be vast as the eastern sea,

Your age enduring as the southern hills.

Let all sing these days of general peace,
And rejoice together in the opening spring.

Now comes the time for holiday robes & caps,
The day when respect & politeness are seen.

May every door have luck & joy,
And every land be blessed with peace.

福	Happiness
申 天	descends from
自	heaven.

The [~~geneal~~] air comes down with riches & honors to this door,

Vernal gales enrich the house with luck in their train.

May times of enduring peace happily prevade the land,
And long-lived people everywhere be found.

Let joy, office, age—three lucky stars—enter this door,
And sons, riches, honors, together bless your gate.

As heaven adds years & months, so do men add age,
As spring fills the sphere, so let joys fill this house.

Clouds involve the phoenix gate of the royal citadel,
And showers & building trees inclose the people's humble houses.

Books brought
Ogilby's Japan, belongs to Nye
Travels of Pinto[27]
Annales of the Daïri[28]
Blue Book, belongs to Parker
Titsingh's Unecdotes
Titsingh's Mariages & Funerailles
Chinese Repository, Vol. Ⅵ, Ⅸ, Ⅹ & odd no$^{s.}$
Meylan, Fischer & Doeff
Belcher's Samarang 2 vs
Halls Loochoo
McLeods voyage of Alceste
Voyage of the Lord Amherst
Krusenstern's Voyage
Lisiansky's voyage
Thumberg's Flora
Japanese Dictionary
Medhurst Vocabulary
Siebold & Rodriguez grammar
Medhurst Chin Dictionary
Williams Vocabulary
Observations on Japan
Missions in Japan
Raffles' Java, Vol. Ⅱ—belongs to Perkins
Collado's Jap Dictionary
Manners of the Japanese
Langsdorff's Voyage
Life of Xavier
Spiritual Songs
Spanish & Japanese Dict
Japan in 1623
Voyage of the Morrison, Vol. Ⅰ
Martin's China, Vol. Ⅰ

San Kofk by Klaproth[29]
Asia Polyglotta
La Perouse, abdgt. of Breton's Japon, 3 vols

Outfit & Expenses connected with the cruise to Japan
Dr.
Cash taken in copper cash 5.00
Cash taken of Ahü on leaving, May 6th 10.00
Cash taken of J.G. Harris for order on
S.W. Bonney fav. JW Jnos[n] 6.00
Cash taken of Chinese Repository sent
to San Francisco 30.00

Dr.
Jan. 11. 1854. John's[30] pay for
4 shirts & writing desk $ 7
Bent's pay for map copying 3
Akin paid from a/c 15
Johnson for Calendars 16
Sarah from Apò's money 0.50
Seal made for Perry 2.50
Dr. Wilson to pay Jno Johnson 2.00
Dr. Smith for Repository 50.00

May 27th

Purser of Powhatan on a/c 30.00
June 1st
Purser of Powhatan on a/c 10.00
June 22nd
Purser of Powhatan on a/c 30.00
June 23rd

Purser of Powhatan on a/c 30.00

———

Hitchocock's money ………….. 240

Less ret[d.] 166

———

74

———

290.00

Passages	16
Lo on a/c	15
Aman a/c	14
Hitchcock	74
Dent	8
On hand	43.25
Outfit	9.75
Sundries, purchases in Jap	110

———

$290

——— Cr. ———

5.70 at Napa

Overplus on brocade	1.50
Furnished Lo on a/c	15.00
Picture of N.Y.	6.00
Plough .90[x] 10 cups 76[x]	1.66
2 hair pins 14[x]	
1 yakate 55	0.69
2 pr trays 40[x] 1 cap 69	1.09
5 fans 15[x]	
2 boxes, 1.11 & 44	1.70
1 teapot 14 10 plates 42	0.56

2.15

1 dress 1.11 1 pillow 7 1.18
sandals 7 cap 90 0.97
Waiter of Lo 2.25

Total 222.34
Shoes 1 8.00
43. 25 Bal. on hand
Passage to Macao Aman & Sundries
at Ningpo & Amoy 15.41 245.75

290.00

Outlay	
Brush & blacking	0.75
Morrison's picture	1.75
Coming ashore	0.25
Postage & boat off	0.33
Little mem$^{o.}$ Books	0.08
Paid Steward	3.00
Washing at Bonin's &c	1.25

------------ Cr. ---------------

Passage to Hongkong $	8
of teacher & boat	1.25
Knife, brushes for teeth	1.50
Wash bowl &c.	2.00
Cash bought in Canton	3.00
To Tarrant for getting names	1.00
To Perry for coins exchange	3.00
For a pair of vases	0.50

For shoes & trowsers	3.00
	5.64
For Japanese books	4.25
For 2 ps cotton cloth	0.64
For crockery, 10 cups	0.75
at Hakodade	
For crape & medicine box of Lo	5.66
For sundries, crockery & med. box	2.00
19.08 at Hakod. May 23d	
For pongee $ 4.16 Crape 4.58	8.74
For 2 trays 16 Glass 17 flask 17	1.04
For 1 Tray 43 roll silk 60 red box 12	2.40
For 1 Inkstand 20 pipe & pouch 25	0.94
For 1 chowchow box 55 weight 5	1.25
For 8 cups, covers & saucers 20	2.16
For stands 8 button silver	0.58
For basket	0.07
For ink — saucers — salt-cellars &c. 1.90	19.08

9.79 Hakod. May 24th

For roll of brocade	0.25
For set of square jars	0.52
For Budhist fish head	0.42
For medicine box	0.52
For books 8 vols.	
4/- coines 4/	1.00
For 2 baskets	0.42
For Ink 2, 20 cups for saki,	
plane .17	0.19
For small set of 3 boxes	0.27
For 2 pipes, 2 razors	0.19

5.24 May 26th

For 2 stone Budhas	2.00
For 10 lackered bowls &	
covers	0.82
For 1 jar 42 1 kappa 65	1.07
For 1 lackered saki cup	
& cover, books, & stand	1.00
For 7 tobacco pouch	35

17.92

Brown lackered slt	
of 4 boxes	5.20
Chowchow box in frame,	
Black	3.00
Chowchow box in frame	
particolored	2.00
Set of five cups,	
thin, red	6.25
Stone image of Budha	0.15

Pictures of people	0.75
Basket 25	
small cups 20	0.45
Small furnace black	0.12
14.49	
Ink 20, books 92	1.12
10 Soy jars	1.00
raised lacker small box	4.86
2 red above & black below	
stands	1.00
1 writing box with stone	1.00
Square stand for cup	1.00
4 stands & 5 cups to each	4.16
Covered red cup	0.35
4.18 May 31st	
Japanese dress	1.78
Tray with fish	1.44
doll 8 tea 25	0.33
5 red trays 42	
5 black 21	0.63
6.42 June 1st	
Changed for Jap. cash	1.00
Two mirrors	4.00
5 cups & covers &	
2 jars blue	1.42
5 cups. Paid Aman	4.00
Inlaid writing box	1.50
Decanter and flower 2/	2.75
Tobacco pouches,	

Pipes & ball	3.18
2 papers tobacco	
& teacups	0.56
4 sets saki cups 3 each	1.25
1 jar 18 4, small cups 2/	0.44
10 cov[d.] green lac[d.]	
cups & cov.	1.44
2 boxes fine porcel[n.]	0.48

bazaar, 21st $15.15

prs of challis or [~~to~~]	3.37
2 trays for Vrooman	0.75
10 sweetmeat cups	0.25
1 perspiration jacket	0.13
lacq[d.] hat	1.12
rain cloak with net	1.40
2 straw cov[d.] boxes	0.75
3 dogs, hairy, toys	0.88
1 soy jar	0.31
lot of hair pins	0.56

39.32 Simoda 23rd June

wooden shrine	0.75
9 cups & 10 lacq[d.]	
bowls of Lo	2.00
Dictionary	1.00
ps of 20 & 9 saucers 9/	4.00
smoking brazier	22.00
red bowl	1.62
green biwk & cover	2.50
two trays	0.75
7 cups thin red lac[r]	1.32
7 cups various sorts	

red lacquer	1.63
3 cups red lacquer	
deep	1.25
5 month pictures	1.00
Japanese coin	1.00
piece of black silk	4.00

Feb. 24th Bay of Yedo

Yesterday the wind blew so hard that there was no such thing as going ashore, nor did any natives come to the ship. I was busy all day in putting the press up, and looking up the various articles belonging to the printing department, which however are so few as to be of little use, especially the assortment of type.

This morning the steamers all weighed anchor to go up the Bay, leaving the Macedonian to wait for the Saratoga. The day was beautiful, and we passed up within a seeing distance of the shore, sounding all the time, & feeling our way till we reached the point attained by the Mississippi last summer, and anchored. The people along shore were much excited by the spectacle, and as soon as we stopped, boats containing parties of men & women came to look at the strange wheeled craft, many of them near enough to get biscuit and other things thrown to them. While two or three were thus pleasing themselves & us, a government boat came shoving into their midst, driving them off with cries, they themselves hastening off in all directions. One or two were overhauled, & one man soundly thrashed with a stick, as a memorial to the others. The commodore was about sending an officer with orders, to be conveyed by Sam

Patch, that if these government boats drove the people off, he would drive them off. However, all sorts of boats were soon out of our reach, but the incident is not of promising augury in respect to the feelings of the government at our coming up the Bay, while it evinces the eagerness of the people. By the evening sun, Yedo was plainly seen over the point in a northerly direction, the city reaching along a hillside, and apparently of great area. Some of the surveying boats went near enough to see the sea wall of the city & its embrasures. This evening many fires are seen here and there, and hundreds of curtains were stretched along the shore, all of which could not be for defenses or troops.

Feb. 25th Off Kanagawa

Capt. Adams came back this morning about 9 o'clock from his visit to Uraga, leaving the Vandalia some way down the Bay. He brought a reply in Dutch & Chinese from the imperial commissioner, signed in the former Hayasi Dai haku kami(or Hayasi, the great counsellor prince, or something like this), but in Chinese, as Lin, member of the imperial council, alone, with no other persons joined with him. At the interview, Capt. A. asked for the cards of the officers he was talking with [~~him~~], but neither of the three were written like the title of the one who replied to the Commodore. This letter acknowledged the propriety of the reference to European & American customs in ambassadors from foreign countries repairing to the capitals of the country they visited, & there delivering their errands at court; but plead its inapplicability to Japan, as the Emperor had decided otherwise, that his commissioner must repair to Uraga, where preparations had been made for the interview;

and concluding by urgently requesting the Commodore to return to that place for this purpose. No alternative was, however, proposed in the paper, in case we held out, such as refusing to see Perry elsewhere, or anything of the decisive nature. A longer letter from our old friend Yezaimon was also brought to Capt. Adams, in which the same things were adduced, no alternative being possible; this last letter was written in a friendly spirit, and indicated, at least, that the Japanese were not prepared yet to break off negotiations in case we refused to go back to Uraga.

Captain Adams said that the place arranged was at the edge of the town, in a narrow place between two hills of no great elevation, one of which had been scarped at considerable outlay of labor to accommodate the buildings erected for the interview, these last being a few rods only from the shore. The buildings were larger than those at Gorihama, and the tables & covered trays seen in some of the rooms, showed that there was to be preparations made for an entertainment on a large scale. The meeting with Capt. Adams was not long, and after delivering the Commodore's letter, turning on the propriety of the squadron returning to Uraga, & the peremptoriness of the siogoun's commands on that point. In reply, the same old reasons were alleged why the ships could not lie there, instancing the bad weather then extant before them all, as an argument patent to all. Tea, sweetmeats, & saki were handed around to all, and the waiters kneeled when presenting the cups. Those who spoke to the prince or chief officer, humbled themselves like slaves, & they were the highest officers who had been on board our ships, where, however, no one kneeled to them. Such abjectness must humiliate the person who does it in his own eyes; or, if it does not, it only shows how deeply it has already abased him. The interview being over all returned to the ship, tho' they thought it not unlikely they would have to stay ashore all night. Yezaimon sent Capt. Adams a small present, parts of it proving the low opinion he entertained of us, or else showing how debased he was himself.

On hearing all these points, and reading these communications, the question of returning to Uraga was discussed, the Commodore still holding to his views not to return down the Bay at any rate. There was a great probability that the Japanese would hold off, but it was quite as important for them to obey the emperor in holding the meeting, as it was to have it [at] Uraga. Of course no one would blame him at Washington or elsewhere for finally going back there rather than lose the treaty; and every country had the right to choose what way it would receive foreign officials: but there was yet no risk of losing what Hayashi had said the emperor had decided to grant, and no precedent could be drawn from European courtesy & reciprocal interchange of diplomats to illustrate one's conduct with a people which ignored all such relations. I approved the decision not to return, tho' I would rather have gone there than risk losing all. Yet I do not at all like the way in which this nation is spoken of by the Commodore & most of the officers, calling them savages, liars, a pack of fools, poor devils, cursing them; and then denying practically all of it by supposing them worth making a treaty with. Truly, what sort of instruments does God work with!

Much to our surprise, Yezaimon and two interpreters, one of whom, Namura Gohadjiro, has lately come into action, & enunciates better than either of the others, came on board. They were received in the rear room, on account of Capt. M^c^Cluney's illness, where tea & toddy and cakes were served as usual. He said he had come to get the answer to the letter brought up by Capt. Adams, & it was promised to be ready by Monday noon. Intimations were given that if they would bring us wood & water, we would pay for it; to which they answered that they could be furnished by bringing them up from Uraga, & hoped our boats would not go ashore to get them. We rejoined that we were not in need of such supplies, and as we know wood & water could both be procured ashore near us, it was needless for them to bring such things from Uraga, and we would not go there to get them. They must themselves have seen two days ago how rough it was at Uraga, & how impossible it would have been to receive supplies from off shore. All this talking occupied some time, during which several things were exhibited, and an india-rubber globe, which Perry made Yezaimon a present of, examined; he was as polite & chatty as usual, & we were glad to see him, & he apparently to see us.

Again the question of going down to Uraga was brought forward, & declined. "Well, then, can you go ashore near here this afternoon, & pick out a suitable place?" said he to Capt. Adams. Thus the whole point was given in, & this was doubtless decided on by the Commissioner at Uraga, as soon as he heard the ships had gone up the bay. The manner in which it was done, showed that Yezaimon was sent up to settle a place for the interview before we got any nearer Yedo; but it came in, during the conversation, as a man gives up a desperate case, by a complete turn-round. A place was pointed out in shore, where he supposed a good spot could be found; and it was decided to go immediately, it being now a quarter of three o'clock.

Capt. Buchanan went with Capt. Adams in another boat, preceded by Yezaimon, and taking a SW. direction, we landed about five miles from the ships, sounding to ascertain the deepest water, at a hamlet below Kanagawa, called Yokuhama. The Commander demands a locality which can be covered by the shipping. A vacant spot of ground was selected near the hamlet, now covered with a promising wheatfield, as suitable for the interview; it was coolly proposed, before reaching this spot, to demolish three or four houses in the village to make room for the new buildings necessary, Yezaimon seeming to think the property of the villager's of not the slightest consideration. He was always spoken to by them on their knees; none of them wearing swords, and showing plainly their low condition by their dress & miserable habitations. The fields were highly cultivated, but the dwellings indicated little thrift, and the village was rendered unsavory by the numerous vats, thatched over to retain urine, compost, & other manuring substances from evaporating, which lined the waysides. Many of the dwellings were built of dried mud & straw supported by cross joists & beams, a few of boards more neat looking than these if no warmer, and the majority of posts and sliding doors. No regularity was observable in the streets or size of the lots, which consequently gave the village the appearance of an incongruous collection of huts and sties, and not nearly so regular and pleasing as the villages around Napa. A few houses were tiled, the ridges being smaller than in China, & imparting a neater look to the roof, which as well as the walls, were white washed white & slate in as pretty manner. The walls of these houses were fully two & a half feet longer at the base than at the eaves. We saw a machine made of two rollers inserted in frame, having each a short screw at their ends working close in each others threads, intended to clean cotton of its seeds, some of which were lying by it. The cotton had a very short staple. Many rude presses were seen to press oil from seeds and others from fishes, now not in use. One loom for weaving mats, a mere frame to stretch the warp on, was observed; but most of the houses were shut up. Hedges of living plants, or more commonly of dried bamboo branches or other trees, surrounded all the yards, and gave a slovenly appearance to the farmstead from the leaves and broken twigs lying on the ground, added to which the farm gear was left scattered in the yard. No windows nor chimneys were seen to admit light into the rooms, [~~and~~] or for smoke to go out. The roofs were nearly the thickness of a foot, made of a sort of reed cultivated for the purpose; a fire breaking out [in] such houses would almost certainly involve all [the] neighbors in its flames. The Camellia trees were [in] full flower, and appeared beautifully when disposed [of] hedges; many trees were just bursting into leaf.

In one part of the village, a large collection of a [one or two words lost] gravestones led us to ask where the people were buried; we were told that the bodies were placed outside of [the] village, and their epitaphs here. Many of the inscriptions were in Chinese, & on one recent one I observed many characters resembling Tibetan, tho' I can hardly think they were so, but rather charms. Near one of the best dwellings was a domestic shrine made with a double door, inclosing the adytum in a box some four feet high. No paint was seen on any building.

The men looked healthy & well fed, but the few women who let us look at them appeared oddly with their shaven eyebrows, and not very tidy; however, the cold weather would induce all to put on whatever clothes their poverty would allow. No animals but cats & dogs were noticed in the hamlet.

Sunday, Feb. 26th

Rev. Mr. Jones held service on board ship today, but did not preach. The crew generally attended, but the marines were paraded on the quarter-deck out of hearing. For the first time on a pleasant day, no Japanese came on board. The aspect of our affairs is now promising, & I cannot but hope that God will hear the prayers offered by his people answered by the success of our Expedition. The peaceful opening of this country will be to this debased, inquiring people a great boon.

Feb. 27th Off Kanagawa

Yezaimon and his friends came aboard, and after considerable explanation and illustration, obtained an imperfect idea of the telegraph, which was put in operation for their enlightenment. So mysterious a principle as the galvanic current requires more previous knowledge of electrical & magnetic powers than these people possess to fully understand this mode of application, even if we were enough acquainted with their language to convey a fair description of the machine to them. However, the result was understood, I think. Yezaimon brought a bushel of wheat done up in a straw bag as a present for Buchanan, who had asked him for a speciman on Saturday evening. After a while, he & his friends went on board the Susquehanna, to see her captain and the working of the machinery while going in to the anchorage of Yokuhama; and every part of the engine was shown which could be, much to their entertainment. The cabin furnished a new sight to Yezaimon, as he [had] no chance to see it last year; the usual variety of spirits was served out, cards exchanged,

& good wishes given & received. Two of the officers were from Yedo, when I told them we must go with these steamers up to that city; they said it could not be, that [there] was not water enough, & the government would [not] allow it: "How can we, who have to come so far," asked I [in] return, "stop short of seeing his majesty?" It is doubt[less] disappointing to the court that we have reached this point, and would have been still more so had Perry only gone higher up, near as he could get. Some of the presents of shell-work from Ye-sima, not far from Kamakura, an island in the Bay of Simoda, belong to Idzu, given by Yezaimon to Capt. Adams were shown, much to our interest; it is said to be manufactured there only, and was really a pretty piece of art. Some of the glass-like wiry byssus of the pinna were also seen, forming part of this present.

I went off to the Susquehanna at Yezaimon's request in his boat. The necessity of removing the official boats from guarding the ships against the people generally visiting them was strongly urged on him. It would be unpleasant to have a collision now, as we are forming a treaty, or trying to do so.

March 1st Off Kanagawa

Yesterday no one came on board in the drizzling rain, which I fear will now continue for many days, as the new moon has come in with a rainy mist. [~~and~~] I was engaged all day on the revision of the Treaty. This evening, Capt. Buchanan gave a dinner to Yezaimon and his friends, which passed off very well, ten of his countrymen sitting down to table with six Americans, for the first time in the experience of each party. The dinner was well served, and the Japanese seemed to enjoy themselves like bon vivants

[French=good livers], drinking healths and joining in the toasts as if they are used to it. Yezaimon proposed the health of the President in return for that of the Emperor, and the health of the Commodore, Captains, & officers of the fleet in return for his own, in all respects acting with perfect propriety. This officer certainly exhibits a breeding & tact in all the novel positions in which he is placed, that reflects great credit on him, and shows the culture of the social parts of the Japanese character. All the guests, except Saboroske, behaved well, but his restless curiosity and impudence led him up & down the room at a great gait;— putting on Capt. B.['s]cap & looking at himself in the glass, hopping behind Yezaimon to take notes, bawling across the table, asking the English name for this thing and that, and making himself conspicuous as a braggart can. Yet his cleverness shines through all his [quarks], even if he did pour out a glass of sweet oil to drinking it for wine. All the guests took parts of the dinner home in their nose papers, wrapping turkey, pie, asparagus, ginger sweetmeats, and other things one after the other; Namura added two spoonfuls of syrup to his ginger, & thrust the parcel in his bosom. Altogether it was a good move, I think, and after dinner they soon returned home at sunset, inviting me to go and spend the night ashore, which however I thought best to decline, on account of the work just now on hand. Before parting one of them sung a song, to which another added the refrain or chorus, but such music! The Japanese can be no better than the Chinese, if such singing pleases their ears.

March 2nd Off Kanagawa

No officials came near the ship today, and the guard-boats which have rowed round the ships to prevent natives coming near us, have disappeared, tho' doubtless the restrictions are as close as ever, given from on shore, as no boats come near us. The draft of the treaty we propose for them to accept is nearly ready, and also Perry's letter to accompany it, a speciman of diplomatic special pleading and foreshortening quite refreshing to a beginner, tho' what is said is well enough, the points which are untouched being the completion of the whole subject. In the evening, I accompanied Capt. Adams on shore to see about the progress of the houses, arrange how to land the escort, and get a walk if we could. There are five buildings, the materials being the same as those employed at Uraga, which have been transported hither. They are to be shingled, and the floors matted; and several rooms like cloisters intimated their supposition that it would be necessary to remain in the buildings some days. They are cheap affairs, and ought to revert to the unlucky owner of the despoiled wheat-field as a compensation for his crop. A flag was fluttering in front, inscribed Go yio 御用 , to intimate that the government had applied its power, and on the limits was another marker called Go-yio chio 御用場 or Arena used for the Emperor. Many villagers came down to see us, but a high officer from Yedo happening to arrive while we were there, the crowd drew off to see him. This magnate was followed on foot by a sword-bearer & shield-bearer, but we happened to be too far off to see him plainly. No such thing as a ramble was possible while so many officials were near, & we soon left.

Friday, March 3rd

Yezaimon sat & drank so much at Buchanan's dinner he was unable to come off to the ship as requested last night. A new & superior interpreter came with Saboroske, named Moriyama Yenoske, who had recently returned from Nagasaki, whence he arrived in 25 days, and hurried on at that. He speaks English well enough to render any other interpreter unnecessary, & thus will assist our intercourse greatly. He inquired for the captain & officers of the Preble, & asked if Ronald McDonald was well, or if we knew him. He examined the machinery, and at last sat down at dinner in the ward room, giving us all a good impression of his education and breeding. Saboroske brought a native map of the Bay & region contiguous, which was copied while he was on board. His principal business was to let us know the Saratoga was off the coast, to bring back a [hammse] for floating, and to arrange respecting watering the ships. He says the houses on shore will not be ready for [some] days yet, so that we shall all have time enough to get ready. I suspect the nearness to Yedo will bring many spectators from thence.

March 4th Off Kanagawa

A party came today for the purpose of bringing an answer from the Saratoga, which vessel anchored this evening. They remained on board almost two hours, drinking & eating, giving me at the same time some practice in talking with them, tho' I got no information from them of any importance. Their chief design was to get something to eat & a glass of toddy, if one might judge of their liking for the refreshments. One of them took drawings of all the parts of a revolver, chiefly by rubbing india ink on a piece of thin paper laid on the things he wished to sketch.

This party afterwards repaired on board the Mississippi and there got some more drink. Mr. Spalding was showing one of them a Prayer-book, & as he turned over the pages, he came to a plate containing a cross drawn prominently; whereupon he dropped the book as if it had been a hot coal. Pity 'tis, that this symbol is associated in their minds with all that is treacherous, dreadful, and forbidden.

Sund[ay], March 5th [Off Kanagawa]

Notwithstanding our request, Yezaimon, Moriyama Yenoske & others came today; it is of little importance to them that it is our Sunday, for we still receive them; they ought to be refused, if the 4th commandment was held in Jewish respect, but what would then be said! Yezaimon had recovered from his dinner the other evening, & appeared in usual health. He said he would come tomorrow, & after examining the telegraph, would return on shore with Capt. Adams to examine the house at Yokuhama. He asked the number of Perry's escort, which was placed at thirty officers, & a guard similar to last year's, but was told that no refreshment need be provided for the guard. The flags we wished to make for doing honor to the siogoun & Commissioner Hayashi were minutely explained to him, & he promised to furnish the diagrams for both; and also a list of the officials & high personages to whom presents ought to be given by us; illustrating both these requests from us by telling him that if he was in America, he would wish to learn such things to avoid blunders. The credentials of Hayashi were also demanded to be brought off for inspection by the Commodore, his own being already in their hands, just as those of the prince of Idzu were shown last year. A mark of confidence in us would be given this year, for they know us better now, in that no Japanese would be marshaled; & we again assured [~~us~~] him that the guard was merely to do honor to the occasion.

A request was made that if any vessels appeared on the coast, as Perry expected some, pilots might be sent off & he informed of their arrival. Yezaimon wished to know how much coal we should annually want, where we wished a port, and what sort of provisions? It was replied that no one could tell how much coal would be needed, but a port on the southern coast, accessible by ships passing on to China or California, where such provisions as they had could be also obtained, would be wanted. He said the most & best coal came from Kiusiu, little from Nippon & none from Sikokf. The Russians were supplied with some, which was pronounced pretty good. Many of these items & requests, especially that relating to the ports needed, were deferred to the Commdore's decision; they were only fishing for answers on the principal points, I think, so that they might frame their replies.

March 6th Off Kanagawa

Yezaimon & his company brought off the copy of the commissioner's Credentials and his emblazonry, as he promised yesterday; the latter was given on one of his excellency's crape overcoats, brought for accuracy. The list of persons to whom presents are due officially consisted of him & his three associates preceded by the six councillors; but what a cloud of obscurity rests over the distribution of these things to them from our utter ignorance of the persons here named!

The day of meeting was fixed for the 8th and after minutely examining the telegraph and the ship, Yezaimon left in Capt. Adam's boat for shore to examine the house & its capabilities. While on board, Sam Patch was brought before him, & questioned a little as to his antecedents, but the poor boy was in such a paroxysm of trepidation that he hardly knew what he did or ought to do. Prostrate on the deck, he murmured some incoherent words, & could not be induced to stand up, so terrified did he become under the stern eye of Yezaimon, who hardly deigned to look at him. I suspect the Japanese stand in more awe, & are more abjectly submissive than even the Chinese, when before their rulers & magnates. The company today was a peculiarly sociable one, & I was talking with them all the time, acquiring words & practice.

In the evening, I made a visit to the Saratoga, where I found the officers much less ill than I was afraid from what I heard of their cases. Mr. Wayne is the most of an invalid, & longs to get home; this homesickness is the attendant of men-of-war much more than I ever supposed, a natural result of the monotonous life led & the constant dwelling on the scenes of home.

March 7th Off Kanagawa

The principal business of today has been arrangement of the presents in due divisions according to the list of officers given to us, separating for the siogoun all those articles intended for him by the government, with others of less value, & distributing to the empress, the six councillors, & the four commissioners, such things as the squadron can furnish.

Yezaimon came about two o'clock to ascertain more particulars respecting the escort and time of starting, & what concerned himself quite as much, to get some of our cake & wine, of which these islanders show an entire belief in. From this, he & Moriyama went to pay Capt. Buchanan a visit.

Wed[nesday], March 8th Off Kanagawa

The Commodore's usual good fortune attended him to-day in a fine, clear day, not overmuch cold either. In the morning, we observed a long line of curtains on the beach, & [~~two~~] [~~rows~~] a row of posts each side of the house on shore extending down to the water, with curtains stretched along, & inclosing the space in front so as to exclude all the view. This rather annoyed the Commodore, since it looked like fencing us in, as had been done at Uraga with boards, which we desired not to be erected; & he sent Capt. Adams & me to have them taken down. In fact these curtains are designed entirely for show, & to do honor to an occasion; but Perry wants honor to be given in his own style or not at all. A fair breeze soon took us ashore, and half a dozen officials came down to the pier the workmen were laying of sand bound up in straw bags, to see what we wanted. A few remarks from them showed that they feared the Commodore was sick, or something else had happened to prevent the meeting. I told them that he expected to be ashore at noon, & we had come to see the place beforehand, the jetty for landing, &c., and suggested that as there would be over 30 boats, the curtains on each side had better be removed to allow more room along the beach, for them to arrange. Instantly, the whole curtain was folded up, the stakes & ropes removed, & a clear beach for landing presented. So the commodore had his way in this, and I think it was a good move, for thus no obstacle was in the way of a view or a ramble; but I put it all on the ground of a small space for boats, and this satisfied them. The rapidity with which the "fortifications" disappeared, greatly amused the people on board ship.

Yezaimon & his party came on board about 10 o'clock to conduct the party on shore, and amused themselves with the sailors & looking at the gay dresses of the marines; as usual, Saboroske was flying about, crying out at the top of his voice from whatever place he happened to be in. The various ships sent their boats first to the flag-ship, & by half past eleven, all of the guard & officers were ashore, the Commodore leaving at noon under a salute from the "Macedonian". On reaching the shore, the band struck up, & passing [~~up~~] thro' the lines of the guard, attended by Kaheyōye, the whole party went up to the reception-hall, where Perry met the five commissioners standing in a row in front of a screen of blue silk; we bowed to them, & the whole then filed around & sat on a bench covered with red cloth, while we were also accommodated on a similar bench opposite, the whole company disposing themselves along two rows, with a low bench before them to serve for a table. Yenoske then separately introduced each commissioner, & a few others, to the Commodore, after which the former retired, each followed by his sword bearer; and a plate of candy was set before each, with tea & fire for smoking. The centre was occupied by a few brasiers on stands, but there was no need of them & little heat in them. Soon after the confectionary & tea had been served to all, the commissioners returned, & invited Perry & his suite to enter a side room, where Com. Lin had [~~him~~] us all seated, and a few compliments, brought out the Emperor's answer to Fillmore's letter, written on a few pages of coarse paper. It acceded to the demands for good usage of shipwrecked sailors, and supplies of provisions for ships needing them, and offered a port for trade, to be chosen by us, and a supply of coal to be there delivered as soon as we needed it. Five years were needed to complete their arrangements for trade at this port, but traffic in articles could be commenced soon.

A Dutch translation was handed in, but the original was not given at this time, as they had no signed copy with them. Our draft of a treaty & explanatory letter were handed to them, and the desirableness of their forming a treaty with us, which would fix our international relations with them on a clear basis; fully dilated on. Notes of several things to be considered were then handed to them & they are to reply in writing. Moriyama was on the floor, shuffling from one side to the other, while these men regarded him with undisturbed countenances, & spoke to him in a very low voice. Yezaimon, Kaheyōye & Tatsunoske were in the room, the latter crouching on hands & knees: what respect can a man have for himself in such a position?

The chief commissioner was an unintellectual looking man, dressed plainly in dark silk. The second is a gross, sleepy looking man, as much unlike a prince as he was a chimney-sweeper; his next in rank taking the shine off all of them by his green trowsers & their gilt emblazonry, he having his coat-of-arms worked on each calf so as to be conspicuous. It was this man who met the party at Uraga, & appeared there much brighter than on this occasion. The fourth & fifth commissioners said almost nothing, and did not present anything attractive; all of them doubtless looked at us as carefully as we at them, regarding us with more interest doubtless, as they had more at stake.

We were entertained by the two princes, while the others went out to look at the papers. Two trays of fish differently dressed, surrounded with boiled seaweed, walnuts, carrots shredded fine, and eggs, were served, with saki, tea, soy & vinegar. As little salt was used as by the Lewchewan cooks, yet the viands were not badly toasted, & I had a fresh supply of the kurumi or walnut seeds, which tasted very pleasantly. A decanter & glasses were brought in, with Madeira wine, which were obtained doubtless of the Dutch. No great outlay was made for today's entertainment, if this was the criterion; but it evidenced good feeling, on the part of the Japanese, and was a vast advance on last summer's meeting.

When the other commissioners returned, they were all invited to dinner; and accepted the invitation finally, as soon as the intimation was given that the machinery would be set agoing for them. The case of the death on board of the Mississippi was then introduced, & a request made that a place be set apart for his interment. First, they wished to know whether the deceased was an officer; then that we take the body to Uraga, whence they would take it to the burial-ground at Nagasaki; this being denied, & Perry proposing Natsu-sima, they raised scruples respecting the [~~property~~] proprietary of the land; and after a deal of backing & filling, agreed to let the body be buried ashore tomorrow, they sending guides to point out the location. All this discussion took up $^3/_4$ of an hour, and allowed the officers outside to see a good deal of the neighborhood, some of them walking a mile or more.

Nothing could be obtained from the commissioners respecting leave to go ashore, & the replies to this & other points in the notes were to be given soon. I have given the leading points in this interview, but the slowness of the intercommunication thro' Dutch too, prolonged it to weariness. While we were inside, the crowd of Japanese outside entertained themselves with the guard, the officers, & the music, and got on very well together. There were about 700 foreigners on shore & lying off. After looking at the long shed for the presents, which required a new roof before they would be safe, Perry & his suite went off. When he landed, salutes were fired in honor of the Emperor & Com. Lin, a mark of respect the latter seemed to understand; these were fired by the boats.

Escorts of Japanese soldiers, crossbow-men, matchlock-men & servants, were standing around the building, but the crowd was never in the way; & both parties mingled freely with each other. The meeting passed off pleasantly in every respect; & towards evening, a dozen boxes of oranges & [~~jars~~] casks of spirits were sent off to the flag-ship for distribution.

March 9th Off Kanagawa

Moriyama & Kaheyōye, the deputies from the commis came about $1^{1}/_{2}$ p.m. to deliver a certified copy of the answer read to us yesterday, & a Dutch translation. It is a mean-looking style to return the answer to the magnificent boxes in which Fillmore's was handed them, tho' this matters little to the contents. These papers were handed to the Commodore, & a short time allowed for dinner, during which some good daguerreotypes of the visitors were taken, & then we went off to the Mississippi, to consult on business. The deputies had Hiraiyama Kenzihrō for their advisor & secretary, but yesterday he acted an equal part with Kaheyōye, & must hold high office. The chief matters settled were the landing of the presents on Monday; the best way of procuring provisions thro' a purveyor, who was to bring them all to one ship, where they would be paid for, weighing coin against coin, equal weight being equal value; and the nature of the presents we wish in exchange for ours. We talked about ports to be opened; the place whence the cannel coal they had brought us was obtained; the desirableness & objections to our going ashore to walk; and need there was on our part for patience in this negotiation, which to them was so novel, & heretofore so opposed to their laws. The princes & commissioners are unacquainted with us & our customs, and much of our success depends on the first steps.

While still in session, the funeral boats returned, Yezaimon coming back with them. The grave was dug near the burial-ground of Yoko-hama; and after Mr. Jones had gone thro' services, a Budhist priest, who had joined the procession, all shaven & shorn, & in a yellow surplice of a fine quality, went thro' his services, having brought his bell & candle, saki, incense-sticks, & all his furniture, to join in this Christian burial. His ritual was much the same as in China; & all present, including over 2000 spectators, regarded it all in quiet interest, somewhat doubting, perhaps, what they would see next. Thus did the U.S. marine Williams occupy his narrow bed within 15 miles of Yedo, where Gongin-sama declared once that no Christian should ever come; yea, that even the God of the Christians should die if he came. Thus are old things passing away in Japan. Mr. Jones thinks he had done a great achievement.

Names of the 6 members of the Imperial Council and the Commissioners appointed to meet Commodore Perry at Yoko-hama.

6 [members of the Imperial Council]

阿部伊勢守　Abe, the prince of Ishi,
in the division of Tokaitu

牧野備前守　Makino, the prince of Bizen
in Sanyuto

松平和泉守　Matsudaiïra, prince of Idzumo,
opposite Ohosaka city

松平伊賀守　Matsudaiïra, Iga's prince; east of Kaga
久世大和守　Kuzhei, prince of Yamato, near Idzumi
内藤紀伊守　Naitō, prince of Ki, in Nankaitu

[the Commissioners]
林大學頭　Hayashi, dai gaku no kami, of the high councillors
井戸對馬守　Ido, Tsusi-sima no kami, prince of islands near Corea
伊澤美作守　Izawa, prince of Mimasaki, in Sanyuto, w. of Miaco
鵜殿民部少輔　Udono, mimbu shiyoyu, Assistant in the Board of Population or Revenue
松崎満太郎　Matsusaki Michitarō

March 10th Off Kanagawa

The answer to the reply delivered by Hayashi has been translated today; & in it, while Perry is pleased that the Japanese government has granted what Fillmore asked for, which was all the Cabinet at Washington expected to obtain, he says that it is by no means all he wants, nor all the President intended, & "will not satisfy his views". The Letter last year asked for one port; now Perry wants five; that desired the Japanese to give assurances of good treatment; now Perry demands them to make a treaty, and threatens them in no obscure terms with a "larger force & more stringent terms and instructions", if they do not comply. The Japanese may be disposed to comply, but they may not. Yet what an inconsistency is here exhibited, and what conclusion can they draw from it, except that we have come on a predatory excursion?

I hardly know just the position in which to place such a [~~double-dealing~~] document as this; but the estimation of its author is not dubious. Perry cares no more for right, for consistency, for his country, than will advance his own aggrandisement & fame, & makes his ambition the test of all his conduct towards the Japanese. Yet if they will, either from fear, from policy, or from inclination to learn & see more of their fellow-men, open their ports, & for once do away with the seclusion system, great good to them will result, their people will be benefited, & the stability even of the state increased, perhaps. Yet I despise such [~~crafty~~] papers as this drawn up this day, & it may defeat its own object: it certainly has lowered the opinion I had of its author.

Yatsnoske came today to see whether the presents would be ready, & to inquire respecting the supplying of water. If he came for wine & cake, he was disappointed. We have given the visitors a large feasting, & it is time they reciprocated it. The arrangements respecting provisions are not very simple, but the supply of such a squadron, where the interpreters are few is likely to be tedious, even if this part of the country has enough — a doubtful matter.

March 11th

Capt. Adams took the papers ashore today, and continued there consulting & arranging with Kaheyoye about the provisions & disposal of the presents. These are now all put up ready for transmission on Monday, and form a large collection, tho' not very valuable. I have had the chief management of their preparation, and the vexatious manner in which Perry can annoy those under him without himself caring for the perplexity he occasions, makes me glad that I never was disciplined to the navy, where undistinguishing obedience is required. One gets into such a heartless way of doing everything, that the whole soul gets callous; praise is never given when a thing is done well, [~~but~~] and scolding, plentifully administered annuls all desire to exert one's self to please a superior.

March 12th Off Kanagawa

The weather during the weeks we have been in this Bay has been delightful, on the whole as healthy, I suppose, as any climate in the world. Today has been cool & clear, (therm. about 42°F) & as bracing as any temperature I ever felt. My health is good and I have enough to do, my situation is not disagreeable, and I am mostly my own master — why should not my heart praise God for all his loving-kindness, so infinitely beyond my deserts, and all the promises given in his dear Son? Mr. Bittinger prayed & read a chapter today, the Commodore having such a tenderness for the crew that he would not keep them on deck in the cold long enough to hear a sermon! He himself attended, but McCluney keeps away. Yet even this slight religious service, which rightfully ought to be held daily, if a crew was properly taught, is made the subject of ridicule and scorn by officers & men, so perverse are they.

March 13th

By eleven o'clock this morning, all the presents distined [sic /destined] for the Emperor and his councillors & the five commissioners were landed or on the beach ready to take ashore. Unfortunately, the day was rainy, and the marines and officers were unable to do more than salute Capt. Abbot as he came ashore, and accompany him into the house, when the former retired to the shed. Most of the presents were landed without injury, & placed under cover, the agricultural implements forming the largest bulk. The engine and telegraph require some preparation to show them.

The presents for the Emperor were as follows:

One $^{1}/_{4}$ size miniature steam-engine, track, tender & car.

Telegraph, with three miles of wire & gutta percha wire.

One Francis' copper Life Boat.

One surf-boat of copper.

Collection of Agricultural Implements.

Audubon's Birds in nine Vols.

National History of the State of New York, 16 Vols.

Annals of Congress, 4 Vols.

Laws & Documents of the State of New York.

Journal of the Senate & Assembly of New York.

Lighthouse Reports, 2 Vols.

Bancroft's History of United States, 4 Vols.

Farmer's Guide, 2 Vols.

One series of U.S. Coast Survey Charts.

Norris' Engineering.

Silver topped dressing-case.

8 yds. Scarlet Broadcloth, & ps. Scarlet Velvet.

Series of U.S. standard yard, gallon, bushel, balances & weights.

Quarter cask of Madeira.

Barrel of Whiskey.

Box of Champagne & Cherry-cordial & Maraschino.

3 10/c boxes of fine Tea.

Maps of several States & 4 large Lithographs.

Telescope & stand in box —

Sheet-iron stove.

An assortment of fine perfumery, about 6 doz.

5 Hall's Rifles, 3 Magnard's Muskets, 12 Cavalry Swords,

6 Artillery Swords, 1 Carbine & 20 Army Pistols in a box.

Catalogue of New York State Library & of Post-Offices.

2 Mail Bags with padlocks.

For the Empress

Flowered silk embroidered dress

Toilet dressing-box gilded

6 doz. Assorted Perfumery

For Commissioner Hayashi

Audubon's Quadrupeds

4 yds. Scarlet Broadcloth — a clock — a stove — a rifle

Set of Chinaware; tea-pot — a revolver & box of powder

2 doz. Assorted Perfumery — 20 galls of whiskey — a sword — 3 box 10/c fine Tea —

Box of champagne — 1 Box of finer tea

For Abe, prince of Ishi, first councilor

One copper Life-boat

Kendall's War in Mexico & Ripley's History of that war

Box of champagne — 3 boxes fine tea — 20 galls Whiskey

1 Clock — 1 Stove — 1 Rifle — 1 Sword — 1 Revolver & powder

2 doz Assored Perfumery.

4 yds. Scarlet Broadcloth

For Makino, prince of Bizen, 2nd Councilor

Lossing's Field Book of Revolution — 10 Galls. Whiskey

Cabinet of Natural History of New York — 1 Lithograph

1 Clock — 1 Revolver — 1 Sword — 1 Rifle — 1 doz Perfumery

For Matsudaiïra, Prince of Idzumi, 3rd Councilor

Owen's Architecture — 12 assorted Perfumery.

View of Washington & plan of the city

1 Clock — 1 Rifle — 1 Sword — 1 Revolver — 10 Galls. Whiskey

For Matsudaiïra, prince of Iga, 4th Councilor

Documentary History of New York

Lithograph of a Steamer

12 Assorted perfumery

1 Clock — 1 sword — 1 rifle — 1 revolver — 10 Galls. Whiskey

For Kuzhei, prince of Yamato, 5th Councilor

Downing's country Houses

View of [~~Georgetown, D.C.~~] San Francisco

9 assorted perfumery

1 Revolver — 1 Clock — 1 rifle — 1 sword — 10

Galls. Whiskey

For Naiïto, prince of Ki, 6th councilor

Owen's Geology of Minnesota & maps

Lithograph of Georgetown, D.L.

10 galls Whiskey

1 Clock — 1 rifle — 1 revolver — 1 sword — 9 assd. perfumery

For Ido, prince of Tsus-sima, 2nd Commissioner

Appleton's Dictionary, 2 Vols.

9 ass'ted perfumery

Lithograph of New Orleans

5 galls. whiskey — box of tea

1 sword — 1 rifle — 1 revolver — 1 clock — box of cherry cordial

For Izawa, prince of Mimasaki, 3rd commisisoner

Model of Life Boat

View of steamer Atlantic

5 Galls. Whiskey

1 rifle — 1 revolver — 1 clock — 1 sword — 9 assd. perfumery

box of cherry cordial — small box of tea

Brass howitzer & two earrings.

For Udono, 4th Commissioner

List of Post Officer — box of tea

Lithograph of Elephant — 9 Assd. Perfumery

1 Rifle — 1 revolver — 1 clock — 5 galls. whiskey

1 sword — box of cherry cordial

For 5th Commissioner

Lithograph of a steamer — 1 revolver — 6 assd. Perfumery

1 clock — 1 sword — 5 galls Whiskey — box of tea & cherry cordial

These things were all arranged in the hall, after the collation of tea & other eatables were over, and Capt. Abbot delivered them in the Commodore's name on the part of the U.S. government, and the Commissioners gave thanks for them; they, however, restrained all expression of interest in them, and really knew almost nothing of what they were. The whole affair passed off very well, & if the sky had not wept so much, it would have been a more interesting "funciaŏ" than that of Wednesday last.

March 14th Off Kanagawa

A boat's load of us went ashore this morning to open out & mark the presents, while others were to exhibit & prepare the agricultural implements, the telegraph, steam-engine, and books. My errand was to open all the books, and with the aid of one of the Japanese, to write the presentation. He declined to break a single seal, and preferred that I should make out another triglott list, which he would hand in. Consequently, I had almost nothing to do, and after luncheon, Dr. Morrow & I slipped out behind the house and reached the nearest hills beyond Yokohama without attracting the notice of any of our officials. Having attained this ridge, we started off into the country, selecting the copses and wooded hillsides as most likely to afford flowers and new plants. We rambled from one woodside to another, crossing fields of wheat & brassica to reach them, and found a few in flower; but we were rather too early, the old grass & leaves not yet being freshened by the coming heats of spring. The wheat was seen in great luxuriance, growing in the richest, blackest soil I ever saw, and cultivated everywhere in [~~rolls~~] rows by drilling. The landscape was beautiful, indicating great fertility and culture; from these hill-tops, few or no houses were seen, no farmsteads or hamlets, but here & there a laborer or woodcutter working solitarily, far away from their homes. There were few birds to be seen, pigeons & crows forming the chief part. The trees were beginning to swell, & in a week or more, the country will begin to assume the hue of summer. We rambled along for several miles, feeling as if we were let out of school, when we approached the seashore, & descried a long village beneath us, and a road leading to it, to which last we descended, avoiding the village. The charming prospect from this elevated point, joined to the idea of its having heretofore been hidden to all foreign eyes, rendered it one not soon to be forgotten. The high degree of tillage showed too, that Japan hereabouts is able to support, & does [~~so~~], a dense population. Our list of plants procured was small, but among them was a kind of fern I never saw before, & perhaps new.

Coming down into the road, we were presently taken in tow by a gay dressed watch-officer, whose guard house lay so as to examine everybody going in & out of the village, and accompanied towards Yokohama. He was a pleasant fellow, & willingly told us everything, stopping as we stopped, & behaving kindly to all he met. The whole population of course sallied out to see us, for we now crossed a large, fertile valley, where every person could see us from all sides, and civilly were they behaved too. No flowers to speak of were seen, except Camellias & Peaches, of which they gave us branches, and thus we went on towards Yokohama, escorted all the way by one & another warden of the paths, everybody being as social and happy at seeing us as possible. I did what talking I could, and asked such questions as I knew how[31].

When near Yokohama, one of our officials came up behind us, puffing & sweating, telling us he had been a long way after us, & rubbing the perspiration from his brow. I begged him not to injure himself by overfatigue, but to help us find some violets on the bankside, which he did, & we soon were merry together. In the village, he procured a cotton gin to show Dr. Morrow its principle of working, and cleaned a few seeds; it consists merely of two [~~wheels~~] rollers working on screws made at their ends, the threads of which interlock. On reaching the house, Yezaimon was waiting for us everybody having been sent for to return on board, and very politely accompanied us to the ship.

March 15th

While Kaheyōye was in conference yesterday with Capt. Adams, a messenger came off in haste on board the Mississippi, to inform him that some of our officers were committing excesses on shore & going off toward Yedo in haste. On this being reported, Perry issued an order for all the officers & men on shore to repair on board instantly, firing a gun to add energy to the command, I suppose, for all who could receive the order could hear the gun. Only three were out of hearing, Bittinger, Morrow & I, and a note was dispatched for the former, who had gone as far as Kawasaki, and had caused all this hubbut among these "insulars". He was overtaken & on receiving the order came back to the ship about 9 o'clock, having been well received by the people at every place he came to. His stories of what he saw are somewhat doubtful, at least until further corroboration; but the walk was an interesting one & showed the good temper of the people & the timidity of the government. In consequence of the order, a guard of four marines was landed this morning, and we all felt like prisoners; the entire squadron is out against poor Bittinger for putting all the officers in quarantine, as there was likelihood of their going ashore in a few days; but I doubt if the Japanese are likely to grant permission, tho' they would not interfere to stop us.

We were busied in arranging the steam-engine, laying the track; & translating the list of presents formed my business. Part of them were carried away today; the emperor's remain longer. Mr. Brown took a few daguerreotypes, & the working of the garden engine amused us all for a time. On coming back in the evening, I had the draft of the treaty the Japanese proposes in return for ours to translate. It is in 8 articles, and proposes to commence a trade at Nagasaki the first of next Japanese year in coal, provisions & fuel, to be paid for in coin; & to open another port in five years after; no permission to be given to go about, and shipwrecked sailors & vessels are guaranteed protection and transmission of themselves & such property as is saved to their countrymen. Concerning trade at Napa & Matsumai, there is no permission, but the phrase is, "we cannot now cavil at it". This would intimate that the latter place was more independent than we had supposed; and perhaps the whole of Yesso is ruled by a tributary prince, as Lewchew is. The treaty is by no means well worded, & leaves many points open, tho' its frames doubtless mean to settle them themselves.

Art. I. When ships of the U.S. come to Nagasaki, they shall be supplied with wood, water, provisions & coal; &

if they lack anything else for their necessities, it shall be supplied them as far as we have it. The time for this going into effect is during the first month of next year; after 5 years, we will open another port for their accommodation. — Note. We may mention that the prices of these things shall be according to those paid by the Dutch & Chinese, & that they shall be exchanged for foreign gold & silver coin, and for no other article.

Art. II. Wherever ships of the U.S. may be thrown or wrecked on our coasts our vessels will assist them & carry them to Nagasaki, & hand t/m [them] over to their countrymen there; whatever articles the shipwrecked men may have preserved, shall likewise be restored. — Note. After the 5 years, when a new port is opened, that which has been saved shall be taken to the new port or to Nagasaki, as is most convenient.

Art. III. As it is not easy to ascertain certainly whether those who may be thrown upon our shores are good men or are pirates, they are not allowed to go walking about those places as they please.

Art. IV. The Dutch & Chinese who dwell at Nagasaki are under old regulations, which cannot suddenly be altered; therefore all Americans resorting there cannot be permitted to go ashore as they please.

Art. V. After the other port is opened, if there be any other sort of articles wanted, or any business which requires to be arranged, there shall be careful deliberation between the parties in order to settle them.

Art. VI. As Lewchew is a distant frontier dependency, the matter of opening a port there cannot at this time be caviled at by us.

Art. VII. As Matsumai is a distant border place likewise, & is ruled by its hereditary prince, the matter of making it a port is also hard to cavil at this time. When the ships of the U.S. come to Nagasaki next spring, this point can be leisurely discussed & arranged.

[the Japanese version of the treaty written in Chinese characters, with four seals; followed by one page of the 條約]

Thursd[ay], March 16th Off Kanagawa

The intended meeting between the Commodore & the Commissioners has been postponed till tomorrow on account of the storm; it is a cheerless place on a rainy day in that rude house. The condition of the common Japanese is not so comfortable as I had anticipated finding it, from what I had read. The villages I passed thro' exhibited evidences of poverty in every form; the houses are slight the utensils scattered around few & rude; the domestic animals few, no hogs, cattle, ducks, geese, or sheep being seen, and only a few chickens, dogs or cats; the people dressed in cotton & in tattered raiment, tho' well fed & healthy looking. The houses are dark, when shut up, and this must prevent a good deal of in-door work in gloomy weather our glazed apartments permit us to do. Temples are common, and gods of stone are numerous, some of them like the Briarean images of the Hindus, others as if only deified men, or deceased-persons whose friends had put them up. The idols of the Budhists were usually seen prominent in these collections, and [~~the~~] Ometo Fuh was inscribed in many places. Tibetan letters were seen in two places, perhaps only the common inscription om mani pudme om, which becomes the more mystical the less is known of it.

We entered a shop for a drink; its contents were

sandals, pattens, vessels containing fish, sauces, and other things, spirits, and an assortment of clothing, the whole not worth $10. I gave a few cash to a girl who brought the water, but our official conductors made her give them back. The people were respectful to these officers, yet not cringing; and probably this custom forms one of the strong bonds to keep the people in subjection.

Friday, March 17th Off Kanagawa

The Commodore left the ship today at one o'clock, and was received on shore by the marines and an escort, with music, and met the four commissioners in the house. The conference was altogether about 3½ hrs., and conducted very pleasantly by the Japanese. The refusal to go to Nagasaki at all was met by the proposal of another port, when Perry mentioned Uraga, and they Shimoda, pointing it out on the map. This place has a fine harbor, & the Commodore agreed to it provisionally, saying that he must first examine its locaiton, & would send the Vandalia & Southampton down there immediately to inspect & survey it. It was surveyed in 1849 by H.B.M. Brig Mariner but no chart of it is in the squadron. Matsumai is to be consulted about, and an answer will be given at the next interview on Thursday, while they can say nothing regarding Lewchew; this therefore seems to settle thc question respecting the political independency of that island of Yedo, whatever may be its relations with Satzuma. Sailors thrown ashore are not to be caged or confined, and to be restrained only after they are found guilty.

Thus most of the objections made to their treaty are likely to be met in a friendly spirit, and I hope nothing will arise to mar the beginning of a new era for them. When we were talking respecting the visit of surveying ships to Shimoda, Kaheyoye inquired if Mr. Bittinger was going, which rather amused us; & when I asked him if he was afraid of him, & this set the Commissioners laughing. He said he was not afraid of him, but he made a great muss.

The oysters today were supplied abundantly, & if it had been a little warmer, the visit would have been very agreeable. The telegraph wire is up a mile, the rail-road will be ready for exhibition on Monday, and the various agricultural implements attract much notice. Today, after Perry had left, a man of elegant manners, & high rank, for everybody went down on their knees wherever he moved, landed & inspected everything with undisguised satisfaction. The commissioners came down from Kanagawa in a large barge, ornamented with banners & official umbrellas, and bearing the Am. flag on the side, a compliment I never heard of the Chinese doing. The boat was prettily painted, & rather a gay thing.

Sat., March 18th   Off Kanagawa

I have spent the whole day on shore, taking a list of the agricultural implements, and assisting in exhibiting them to the people around us, many of whom appeared interested in their manipulations. The most of these machines are far too expensive & complicated, I fear, for the majority of the agriculturalists & gardeners of Japan. The operations of the tillers of the soil, here as in China, are on too small a scale for them to afford the cost, and human labor for these same too abundant to need, such implements; & it will take much time to introduce them. The power of machinery, however, can find large fields for its exercise in these remote regions, when once it is allowed full play.

The day passed rather tediously, as I had not much to do, and the knowledge of the language is too limited yet to enable me to talk readily. I had a good opportunity to tell a considerable number of the spectators something about the resurrection, a matter totally new to them, and which struck them as wanting much evidence to lead one to believe it. During a walk to the marine's grave, we saw a few new things, among which the extensive use of charms at graves written in Tibetan & Chinese characters was one. None of our friends knew what the former meant. Many new guard houses have been placed in Yokohama since we came, some of whom [sic / which] are filled with persons bearing the coat of arms of the prince of Sinano.

Many of our visitors today are new, and I learn that several of those formerly here have been relieved by a second set, the others having gone to Uraga, among whom is Yezaimon & Saboroske.

Sund[ay], March 19th Off Kanagawa

With the disregard of the Sabbath usual in this fleet, the Southsampton was kept coaling during all night and most of the forenoon in order to get her ready to leave for Shimoda with the Vandalia. The Supply came up the Bay this morning, disappointing more than she satisfied when her letter bag was distributed. Mr. Jones held service on board the flag-ship, Capt. McCluney as before declining to attend. It is a matter of gratitude to hear of the welfare of dear friends, and get letters in Japan from Utica to Dec. 1st. Mr. Contee's letter describing the landing of last year has been the chief sport for the fleet since the Supply came in.

March 20th

Many changes in the officers of the squadron, ensuent on Dr. Gambrill's death & the return of the Saratoga with several invalids; Dr. Wheelwright & Mid. Stockton leave this ship. The Japanese came aboard twice yesterday, Isaboro being now the chief spokesman in place of Yezaimon, & an inferior man in all respects. Today, I have been ashore all day, and as if I was known now, no Japanese interpreter came to the house the whole time. This practice, of course, is just what I want, troublesome as the impertinent & reiterated questioning sometimes becomes. There were very few visitors today, but many questions when the railcar will be in readiness to move. Some new plants were collected in a short walk, and shells, but we are a month too early for botany.

March 21st Off Kanagawa

Dr. Morrow & I went off this morning on a search for an appropriate place to exhibit his hydrostatic ram, but after rambling two hours along the base of the hills back of the village, we returned unsuccessful. The officials who accompanied us were not much pleased with the tramp thro' tangled underbush and boggy paths, for their straw sandals are ill fitted for getting over rough places. The season is not yet advanced enough to make these rambles pay in botanizing. The locomotive and tender were started on their circuit today, and went scudding round & round the circus like a Shetland pony, to the great pleasure of every epectator. The Japanese are, I think, more pleased with this thing, than anything else we have given them.

[Wed.], March 22nd Off Kanagawa

Another unsuccessful search for a proper place to exhibit the hydrostatic ram; but we came across some petrifications, in the rocks at the base of a cliff, and procured several specimens. The rock was in site 150 feet above water mark, in a friable conglomerate, colored with iron. Some plants were dug up that promise something. Our companions today were not well disposed to an extension of the walk but I managed to keep them in good humor, especially on the matter of procuring a couple of ducks we saw in yard.

A large party came today from Yedo & Kanagawa to see the locomotive and telegraph. We managed to communicate thro' Namura's aid, by writing the sounds in Japanese & sending them literally. It satisfied them, however, and all appeared to understand the idea, tho' not the mode of its operation. This party of people were not a whit superior to any of the previous companies of visitors we have had, & I know not that they were of any higher rank. On reaching the ship in the evening, we found that it had been agreed on to send the revised articles of the Treaty on board tomorrow, & deliver the return presents on Friday.

March 23rd Off Kanagawa

Hiraiyama Kenzhirō came off this morning with the following paper:

[The Hakodate treaty written in Chinese, signed by Hayashi and three other Commissioners, with its English translation; only the latter is given here, as follows:]

"Ships of your nation passing by, and being in want

of provisions, fuel and water, are permitted to procure them at the port of Hokodade, which we desire may be regarded as consonant with the desire expressed in the letter received from you: But as it is a distant place, and time will be necessary to prepare and settle everything there, it is arranged that [in] the 7th month of our next year (Sept. 6th to Oct. 5th) be the date for opening the port.

"Kayei, 7th year, 2nd month, 25th day. (March 23rd 1854.)

"Hayashi, Dai-gaku no kami
"Ido, Tsu-sima no kami
"Izawa, Mimasaki no kami
"Udono, mimbu shiyoyu"

This gives permission to our whalers to repair to the port of Hakodade near Matsmai for supplies, and the time appointed for opening it will probably be as soon as arrangements can be made. Whether it will prove as good a place for furnishing these ships with supplies remains to be seen after a few experiments have been made. It is probably a small & unimportant place now, and time will be required to attract traders & provisions there.

Frid[ay], March 24th Off Kanagawa

The Susquehanna started for Hongkong early this morning, much to the regret of most of her officers, especially the Captain, who are thus disappointed in seeing more of the country of which they will be expected to have learned almost everything; and what annoys them still more, they are unable to get any articles of rarity of Japanese manufacture, or see what is more to be seen of their customs. The Commodore reached the shore in his barge about noon, the four commissioners having been there some two hours before him. On reaching the hall, we found the return presents from their government spread out on the mats, lying in pretty pine trays, and making a pretty show in consequence, far more so than ours did, done up as the most of them were in brown paper & rough boxes. Some of the pieces of lacquered-ware in raised gold figures were beautiful, and the silks were rather fine, especially the heavy crapes; the patterns of these last were quite unlike anything now made elsewhere. The list exhibit the variety.

1st From the Japanese government to U.S.A.

1 gold lacqd. writing-table, writing apparatus, paper box, & book-case, 4 pieces—1 bronze cow-shaped censer, with a silver flower on top—1 set plates or trays—1 bouquet-holder & stand—2 braziers for charcoal—10 ps. ea[ch], white & red pongee, & 5 each figd. & dyed crape.

2nd From Hayashi to U.S.A. Government

1 lacqd. writing apparatus & paper-box—1 box paper, of flowered papper, & 5 of stamped note paper—4 boxes assorted 100 kinds sea shells—1 box holding a branch of coral & a silver feather—1 lacqd. chowchow box—1 box set of 3 goblets—7 boxes cups, spoons & goblets cut from couch shells.

3rd From Ido, prince of Tsus-sima

2 boxes 4 lacqd. waiters—20 paper umbrellas—30 coir brooms.

4th From Izawa, prince of Mimasaki

1 ps. ea[ch], red & white pongee, 13 dolls, box of woven bamboo articles, & 2 bamboo stands.

5th From Udono, member of Revenue Board

3 ps. striped crape, 20 porcelain cups, & 10 jars of soy.

6th From Matsusaki

3 boxes porcelain cups, 1 box fig$^{d.}$ matting & 35 bdls. oak charcoal.

7th From Abe, 1st Councilor

15 ps. striped figured pongee or taffeta.

8th—12th From the other six councilors

10 ps. striped figured pongee, from each councilor.

13th From Emperor to Com. Perry

1 lacq$^{d.}$ writing apparatus & paper box, 3 ps. red & 2 ps. white pongee, 2 ps. flowered & 3 ps. fig$^{d.}$ dyed crape.

14th From Commissioners to Capt. Adams

3 ps. red pongee, 2 of dyed fig$^{d.}$ Crape & 10 sets cups & covers.

15th From Commissioners to Perry, Williams[32] & Portman, each

2 ps. red pongee, 2 of dyed fig$^{d.}$ crape & 10 sets cups & covers

16th From Commissioners to Draper, Danby, Gay, Williams [33] & Morrow

1 ps. red dyed figured crape & 10 lacq$^{d.}$ cups & covers.

17th From Emperor to squadron

200 bundles of rice & 300 chickens. Each bundle

contained five Chinese pecks or 斗 <u>tau</u>.

There are in all 132 ps. of silk. Everything was brought off & except the chickens, are all to be sent to Washington.

After the exhibition of the presents, the commissioners invited Perry out in front, & soon 90 naked <u>rikozhi</u>[力士] or atheletoe, paraded in front, to show their brawn by carrying the bundles of rice in various ways; some, two on their heads, others, one in their teeth, at the end of their arms, or on their back. These fellows are trained to such feasts, and were all stout limbed men; the biggest stripped to let Perry punch him in his paunch. They were brought to this village from Yedo, and we regarded it as a good sign that the Commissioners should take some pains to amuse us. From this spot, the company repaired to the rail-road, where the locomotive was soon ready to run its race around the ring, a spectacle which interested the natives greatly. On returning to the house, the company was seated facing the inner yard, where the strongest of the athletoe were brought forward to exhibit their prowess. First, the whole body of them stood in a circle, and went thro' a sort of drill or manual, slapping their breasts, rubbing their hands, arm-pits, and knees, with other motions, after which they marched off. A second company, dressed a little with long fancy aprons, then circled the ring, going thro' with similar motions. The match then began, two and two coming into the ring. First, squatting on their feet, opposite each other, the two began to rub themselves with dirt on the psalms & arm-pits, and then advanced to the centre in a steady step. Here, each stretched out one leg after the other, holding his knee with a close grip, and planting his foot in the earth with a heavy groan or grunt, several times, again rubbing his hands in the gravel like a bull pawing the earth. All this took up a minute or more, and then each seizing the other's shoulders, endeavored to push his antagonist over; one butted his head with all his force against the other's breast, while that one [~~did~~] only tried to throw him by turning his body, & generally succeeded in doing so, he coming to the ground with a thump that showed the force exerted. In only one case was there anything like wrestling. All the men were heavy, and seemed strong too; the biggest remained victor. Some of them rushed up, screaming like mad, but these generally proved to be weaklings. It was a curious, barbaric, spectable, reminding one of the old gladiators. Indeed, there was a curious mélange [French=mixture] today here, a junction of the east & west, rail-roads & telegraph, boxers and educated athletoe, epaulettes and uniforms, shaven pates and night-gown soldiers with muskets and drilling in close array, soldiers with petticoats, sandals, two swords, and all in disorder, like a crowd, all these things, and many other things exhibiting the difference between our civilization & usages and those of this secluded, pagan people.

The interview lasted two or three hours; at the close of it Comr. Lin gave Perry two swords, three matchlocks & 2 sets of coins. All the high officers seemed in good spirits, and every body left for the ships much amused with the day's sh[owing] how desirable that our opening intercourse may produce different results, calculated to elevate & purity the Japanese, so that they may learn the real source of our superority in the momentous truths of the Bible.

Sat[urday], March 25th Off Kanagawa

M. Yenosike, Isaboro, Kenzhirō, & others came to see Perry to-day, while we were all hurried here & there to pack & mark the presents recd. yesterday. They wished him to defer his visit to Matsumai for 100 days, but he refused to do so more than 50; they said interpreters must go there viâ Yedo, & the dialect differs so much there that I cannot understand them. The Japanese are unwilling to allow consuls, as they say the governor & interpreter can manage all things with the captain of the ship. The discussion respecting trade after the treaty, walking about, furnishing coal, and the immediate opening of Simoda, was on the whole favorable. In the course of the interview, it came out that owing to Pellew's foray into Nagasaki harbor, & the suicides of the native officials, great fear was entertained of the designs and violence of the English. It was agreed to-day that a tariff of prices of merchandize, coal, provisions, & other things be made out, for the purpose of informing our people that ships may know what they are to pay & what they can get in Japan.

[Sunday], March 26th

For once we have had a quiet Sabbath, there being little or nothing doing. A slight rain prevented service on board the Mississippi, and also kept us all pretty much in-doors. I spent a part of the day reading M'Cosh's book, and found much instruction from it. The views therein presented strike me as embracing a fuller view of the ways of God to man than in any treatise I have ever read. Blessed are those who understand those ways to their own salvation.

Mond[ay], March 27th Off Kanagawa

There was nothing done in the ships today but make preparation for the entertainment given her to Hayashi & his colleagues, with other officials; the five former were provided for in the cabin, and about sixty came to the tables on deck. Good humor prevailed and the whole appeared to be gratified.

The commissioners first went to the Macedonian, where they saw an exhibition of the manner of training, loading, & firing great guns, and all the other evolutions of ship's company at general quarters. When this was over, they left that ship under a salute, and were received by Com. Perry on the quarter-deck, and conducted over the ship, including the engine, which was put in motion for their entertainment. It greatly surprised them, and apparently bewildered some of them. Dinner was now ready, and above & below, all prepared to taste the good things provided for them. Capts. Abbot, Lee, Adams, & Walker assisted the Commodore, and they furnished the Japanese with a taste of everything on the table, sipping wine, tasting meats, preserves, pastry & other rarities, until they were all very well satisfied. I managed to tell them the names of nearly or quite everything, which also seemed to increase the interest in the feast. But the appearance of four large cakes, each having a miniature flag with the coats of arms of the four commissioners on it stuck in it, was the best hit; they received the compliment as a well timed one.

About $5^1/_2$ o'clock, all went forward & listened to a performance of singing & dancing by the minstrels, until it was too late for them to stay longer; this exhibition was a source of great merriment to them & every one present, for the acting was excellent. About 200 Japanese altogether were on board, and the day passed off without accident, & to the gratification of everybody.

Tuesd[ay], March 28th Off Kanagawa

At noon, the commodore met the Japanese commissioners on shore and discussed some of he points he had drawn up, including those which had been accepted. That for opening Simonda as soon as the Treaty was signed was objected to so strongly that a compromise was agreed upon, which amounted to deferring [~~any~~] all trade there, except for this squadron, until the President had promulgated his orders and notices that the place was available. Another hesitancy was seen in the limits to which Americans might go from the ports, and it was at last limited to seven Japanese miles, and a man was to be back the same day. "Temporarily" was also inserted before the word "residing", in this article, as they did not see the use of putting residents on shore there. A good deal of discussion of a friendly nature was carried on today upon several points of the treaty, & all its points & articles were settled. How much has been gained over what I expected last spring, when I was asked to come here; how thankful ought we all to be that no collision has taken place.

March 29th Off Kanagawa

Doing up specimens of American coins, and preparing articles of the Treaty all the morning, which Yenoske cavilled at when he came in the afternoon, accepting some and altering or rejecting others. All the management of the Treaty seems to have been transferred to his hands by the commissioners, for Kenzhiro & others with him said almost nothing. In all these consultations, Yenoske seems to possess the decisive authority, and is pretty well fitted for it. Objection was made to the distance allowed for rambles,

& the point was conceded for Simoda, starting from a small island in the harbor, & for Hakodate, [~~at~~] [~~a~~] [~~small~~] when the Commodore has been there. A curious objection was made to the ratification of the treaty, as the emperor needed only to approve what his commissioners had done, and then it would be evermore obeyed. This people seem to be bred into a full idea of the "right of kings divine, to do just as they have a mind", and to liberate them from such a thraldom can only be brought about by the Gospel. The discussion of one point and another, the appointment of consuls, the opening of the port of Simoda, & the distance to which Americans can ramble there, prolonged the interview till dark. During the afternoon, Kenzhiro wrote the following as either expressing his dislike or predilections:

"In the vast expanse of the world's extent, are not all the tender children of the 天帝 Heavenly Ruler? Among them, courtesy, good-faith, kindness & justice ought to rule, as they do among own brothers; but if, covetous of gain, things are carried to an extreme, all ought to be ashamed of it & not speak thereof: yet to discourse of warlike affairs and the necessary modes of commotion, slaughter, & battle, is not unworthy of continual talk and research." If he alluded to the deliberations then going on, it was a hint that we were rather [quiddling].

Thursday, March 30th Off Kanagawa

The same party came early today, and after going over their Dutch version, and making one from it & Mr. Portman's, all the articles, 12 in number, were agreed to, some other points being put into a supplementary letter, one of which was that Simoda is not actually to be opened till next autumn; and another respecting consuls. Yenoske, in return for all that Perry had given him, brought a box of sweetmeats; and it was a pretty box indeed. In return he took away a box of Lowell cottons, and also the presents for Kaheyoye; and a promise of a brass howitzer for the two princes, Izawa & Ido, for which they have been asking again & again. It is not for want of cumshaws[tip] to the Japanese that we shall fail of making a treaty, especially drinkables of all sorts; tho' I suppose this is the way to do such negotiations the world over.

Frid[ay], March 31st 1854. Off Kanagawa

Last evening, Kenzhiro came about 8 o'clock with the Chinese version of the treaty done from the Japanese, and after some alterations and the correction of one important error respecting the distance allowed for rambling at Simoda, the whole was agreed upon. This morning a fair copy was made, and about 12¾ o'clcok, the Commodore left the ship. On meeting the Japanese commissioners, they exhibited three copies of the Japanese version, & one each of the Dutch & Chinese, while we had three copies of the English & one each of the Dutch & Chinese.

They first [~~exhibited~~] opened theirs at the seals, to show the rubrics attached to the name of each commissioner, instead of a seal, & then the Commodore signed the three English copies in their presence. The two copies of the Dutch version were then compared & found to be the same, when they were exchanged, one being signed by Yenoske, the other by Mr. Portman. After this, the Chinese copies were compared, and one character erased in one of them, but when I wished them to sign their copy & date it, a difficulty arose, for they wished only to date in Kayei's name & year, while I required both theirs & ours, as in the Dutch. They declined to write the characters for "our Lord Jesus Christ", and the Commodore allowed the omission, after which they dated it, & Matsusake Michitarō signed it with his rubic; and I signed the other & gave it in exchange. Thus was completed the negotiations & signing of the Treaty of Kanagawa, the first one ever made by the Japanese. Long may they rejoice over the blessings it will bring them, and may the Disposer of Nations & events make it the opening whereby His great Name may be declared unto them. After so many years of seclusion, he has inclined them to listen to this application to loosen the strictness of their laws, and I sincerely hope they will never have occasion to repent of the privileges granted on this day.

The Treaty being signed, a dinner was brought us, tho' it hardly came up to our expectations. The first course was composed of tea, candy tied in knots & sponge-cake. The second of raw oysters, mushroom soup, boiled pear, eggs pressed together after boiling into cakes & then cut into strips, seaweed cooked with sugar, raw ginger, boiled walnuts, and mushrooms, hot & cold saki served as occasion required. The third of boiled bream, large crawfish, shrimp, sliced fish, bean soup with greens, seaweed in fine threads, greens, boiled bamboo & onions, with the long yam, a vegetable I never saw before. The fourth of fish soup, taro, blancmange, with the word shau or longevity on it in a cypher in red, boiled chestnuts, and one or two other unknown matters. As a whole, it was not equal to the dinners given at Shui, and would doubtless have been better served at Yedo or even Kanagawa.

Dinner being over, a long discussion ensued respecting the visit to Yedo, to which the Japanese made many objections, and requested the Commodore as a personal favor not to go up the Bay; but he told them it must be done as the President had ordered it, even if they did not let him go ashore. It ended by the closing further useless alteration of arguments, & each wishing the other good-bye. Commr. Lin said it was the firm determination of the Japanese never to open the port or Bay of Yedo to foreign ships. At leaving, Capt. Adams handed over the list of the presents still on hand.

Sat[urday], April 1st Off Kanagawa

The list of agricultural instruments & seeds were given to the interpreters this morning, and a number of carpenters were ready to begin to pack them up, as well as the telegraph & locomotive. This being done as far as we had anything to say, Dr. Morrow & I started off to collect plants, tho' the slight frost & cold weather lately had rather retarded than hastened their development. We went up the creek & crossed the bridge, where we saw a fellow throwing a net, in which came up a fine surmullet, a silure, & a sort of perch, but he had no means of carrying them off. Proceeding northward across the valley, we reached the hill, and went onward for about three miles, finding little to repay us, but much to see. At one farm house, we procured a little cotton seed, while no one had seen the cotton growing hitherto. In the next valley, we reached the high-road leading westward from Yedo, and came into the village of Hodangya, stretching along both sides of it for a mile. The people were all abroad, and all pleased to see the foreigners, as we were to see them. The shops were low buildings, with nearly the whole front open, displaying only the common necessaries of life. On one sign we noticed the name Vroum von Metter in Roman capitals, and on another the efficacy of a medicine introduced by the Dutch from abroad was extolled. A few two-storied houses, with the gable ends to the streets, seemed to be the dwellings of the better sort; their window-blinds were made of 2 in. plank trebled; some windows were grated. A covered way stretched along the whole street, but not so as to protect foot-passengers from the rain; it was merely a shelter for the individual householder. The road was nearly a macadanized one; a few pack-horses were seen but no vehicles, & almost no animals. The crowd gave way as we went on, everyone preserving the utmost order; among them the women, with their black teeth, looked the more repulsive the more they laughed; and three or four naked fellows who had run out from their work, looked oddly amid the dressed crowd. As a whole, the line of shops & houses did not equal a similar row in China, and the people were not, I thought, as large on the average. Dr. Morrow & I were almost a head above them.

A copy of the Treaty of Kanagawa

The United States of America & the Empire of Japan, desiring to establish firm, lasting, & sincere friendship between the two nations, have resolved to fix in a manner clear & positive, by means of a Treaty or general convention of peace & amity, the rules which shall in future be mutually observed in the intercourse of their respective countries; for which most desirable object, the President of the U.S. has conferred full powers on his commissioner Mathew Calbraith Perry, special ambassador of the U.S. to Japan; the august Sovereign of Japan has given similar full powers to his commissioners, Hayashi, Dai-gaku no kami, Ido, prince of Tsus-sima, Izawa, prince of Mimasaki, & Udono, member of the Board of Revenue. And the said commissioners, after having exchanged their said full powers & duly considered the premises, have agreed to the following article.

I. There shall be a perfect, permanent & universal peace, & a sincere & cordial amity between the U.S. of A. on the one part, & the empire of Japan on the other part, & between their people respectively, without exception of persons or places.

II. The port of Simoda in the principality of Idzu, & port of Hakodade in the principality of Matsmai, are granted by the Japanese as ports for the reception of Amn. ships, where they can be supplied with wood, water, provisions & coal, & other articles their necessities may require, as far as the Japanese have them. The time for opening the first-named port is immediately on signing this Treaty; the last named port is to be opened immediately after the same day in the ensuing Japanese year. — Note. A tariff of prices shall be given by the Japanese officers of the things which they can furnish, payment for which shall be made in gold & silver coin.

III. Whenever ships of the U.S. are thrown or wrecked on the coats of Japan, the Japanese vessels will assist them & carry their crews to Simoda or Hakodade, [&] hand them over to their countrymen appointted to receive them; whatever articles the shipwrecked men may have preserved, shall likewise be restored; & the expences incurred in the rescue & support of Americans & Japanese who may thus be thrown upon the shores of either nation are not to be refunded.

IV. Those shipwrecked persons & other citizens of the U.S. shall be free as in other countires, & not subjected to confinement, but shall be amenable to just laws.

V. Shipwrecked men & other citizens of the U.S. temporarily living at Simoda & Hakodade, shall not be subject to such restrictions & confinement as the Dutch and

Chinese are at Nagasaki; but shall be free at Simoda to go where they please within the limits of 7 Japanese ri or miles from a small island in the habor of Simoda, marked in the accompanying chart hereto appended; & shall be free to in like manner to go where they please at Hakodade, within limits to be defined after the visit of the U.S. Squadron to that place.

VI. If there be any other sort of goods wanted, or any business which shall require to be arranged, there shall be careful deliberation between the parties in order to settle such matters.

VII. It is agreed that ships of the U.S. resorting to the ports open to them, shall be permitted to exchange gold & silver coin & articles of goods for other articles of goods, under such regulations as shall be temporarily established by the Jap. govt. for that purpose. It is stipulated, however, that the ships of U.S. shall be permitted to carry away whatever articles they may be unwilling to exchange.

VIII. Wood, water, provisions, coal, & goods required, shall only be procured thro' the agency of Jap. officers appointed for that purpose, & in no other manner.

IX. It is agreed that if at any future day, the govt. of Japan shall grant to any other nation or nations, privileges & advantages which are not herein granted to the U.S. & the citizens thereof, that these same privileges & advantages shall be granted likewise to the U.S. & to the citizens thereof, without any consultation or delay.

X. Ships of the U.S. shall be permitted to resort to no other ports in Japan, but Simoda & Hakodade, unless in distress, or forced by stress of weather.

XI. There shall be appointed by the govt. of the U.S. consuls or agents to reside in Simoda at any time after the expiration of 18 mos. from the date of the sign[ing] of this

Treaty, provided that either of the two govts. de[ems] such arrangement necessary.

XII. The present convention having been concluded & duly signed shall be obligatory & faithfully observed by the U.S.A. & Japan, & by the citizens & subjects of each respective power & it is to be ratified & approved by the Prest. of the U.S. by & with the advice & consent of the Senate thereof, & by the august sovereign of Japan; & the ratification[s] shall be exchanged within 18 mos. from the date of the signature thereof, or sooner if possible.

In faith whereof, we, the respective plenipotentiaries of the U.S. of A. & the empire of Japan, aforesaid, have signed & sealed these presents. Done at Kanagawa, March 31st, 1854, & Kayei 7th year 3rd month & 3rd day.

[the Japanese version of the Treaty written in Chinese characters; 3 pages]

[Sunday], April 2nd Off Kanagawa

Mr. Jones did not come aboard the Powhatan today, but I had opportunity to go to the Mississippi. Notwithstanding our repeated requests, a party of Japanese came aboard today and remained drinking & talking most of the afternoon. The interpreters have doubtless learned at Nagasaki that the Sabbath is only a longer holiday & nothing of a holy day; & doubt not they will find it so here too. They brought Perry five pieces of crape today in return for the pistols, cloth, & wine he has given them.

Mond[ay], April 3rd

It [~~blowed~~] blew a gale all day, preventing boats going fr[om] one ship to the other, and affording time to write to

U.S. by the Saratoga. I sent letters to Dr. Anderson advising a delay of two or three years before sending misisonaries out to Japan; to Mr. Bidwell, Mrs. Graham, J.R. Bartlett requesting Hanna's Chalmers or Grote's Greece in exchange of books sent him long since, and to Mr. Merwin, & the N.Y. Observer.

April 4th

The Saratoga left this morning, carrying some invalids & Capt. Adams with his Treaty. I have hardly ever been so affected by any music as I was today by the Mississippi's band playing Home; sweet home, as the Saratoga passed her; it brought tears to some eyes. I fear Lieuts. Wayne & Randolph will not live to see their families, & they show no preparedness to meet God.

Wed[nesday], April 5th Off Kanagawa

A stormy, rainy day, preventing every one going ashore, and making it disagreeable on board; indeed, a man-of-war in a rainy day has been few comforts or redeeming points, everybody looks as if they suffered rather than existed, enduring rather living, until it is over and the sky clears up. Yenoske & his party however came off, bringing several boxes of chinaware for Com. Perry and others. I was much amused at the different ways in which one & another of his party protected themselves from the rain, some with paper, some with felt, some with straw, some with varnished cotton, all effectually protecting themselves too, with varnished caps & umbrellas.

Thursd[ay] April 6th Off Kanagawa

Com. Perry & a party landed today for a walk. The howitzer and its two carriages & ten boxes of tea, were taken ashore as the last presents to be made the officers here. The gun will doubless be regarded as a great prize; the first question, was where is the powder & shot, and "let us see you fire it off". I suppose the Japanese will soon begin to cast others like it, and think themselves able to resist foreign aggression as soon as they have made guns. After a few cups of tea had been served, the party started, going towards the old telegraph house, and then into a small mia or [~~ancestral~~] Buddhist temple, having three images, and some tablets. The chief image seemed to be cut out [~~off~~] of the root of a tree. The inscriptions were all in Chinese, but there was no time to get their explanation.

From this, we struck across the rice fields along the dyke, & ascended the hills west of Yokohama, and down into a pretty dell, where we rested in a small [~~house~~] temple for some time. It was a charming spot, and the camellias, peaches and plums, all in full flower, gave it a gay appearance; while the delightful temperature made everybody feel happy. The people living here came out to see the foreigners, but our official escort repelled them to a distance. We saw the tea plant growing in this nook, the first row of it I have noticed.

From this, the Commodore returned to the village, and paid the headman a visit, as I had suggested to him, to conduct us round that way. In this yard was a curious pine-tree, the trunk of which was about 4 feet high, & the top spread out like an umbrella, 20 feet more in diameter; it was the result of 30 years' labor & culture, and was in a healthy state, full of flowers. It was not so large as one we saw last Saturday near a village up the valley, that being on the loamy bank of a stream. Besides this, our host had a fir tree bearing several branches of pine grafted in, which he evidently took some pride in. During the visit, his wife & daughter came out, one bringing his grandson, and making themselves part of the party. It was instructive to see how utterly regardless of the man & his family, Yenoske & his fellows all acted, sitting on the mats smoking, & laughing away themselves. I suspect the lower ranks of life in Japan are kept from rising by an iron hand; and yet how totally unprepared they are for asserting their rights is too plain to every one. This man has been obliged, probably, to accommodate several officers since we came; and perhaps much of the cost of entertaining us & them has fallen on the village. We left and returned on board altogether, leaving nothing behind at the house.

To our surprise our host of this morning came aboard the "Powhatan" about $8^1/_2$ o'clock, bringing with him a fan and a dozen sheets of paper to get my teacher to write him some autographs. He had heard that we were soon to leave, & this was his only way to see the ship. We gave him a few presents, and he departed mightily pleased with his reception; he is a general favorite, especially with Dr. Morrow & me, whom he has accompanied most cheerfully in many a long ramble. I hope he may be able to keep what he took with him, for he told us that he had been forced to give up some of the seed formerly given him.

April 8th, Off kanagawa

Yesterday was a rainy, cold day, and the quiet of ships un[der] the blasts which now & then swept by us, showed the excellence of the anchorage, and the security ships can expect in this place. A heavy storm doubtless was felt on the coasts; but this morning opened clear and invigorating respecting everybody by the [~~clear~~] bright sunshine. Mt. Fusi and all the high land at its base, was covered with snow, showing the little advance yet made in the coming on of spring compared with what might be inferred from the vegetation along shore. Some of it had disappeared before night. The water has been all now delivered to the ships, for which the Japanese will take no pay. No provisions have been brought us for many days, & many a chicken & duck came to an untimely end by the cold last night, so that it is about time to be moving. In fact, our official purveyors have given us very few eatables, and not exerted themselves to supply us with what could be obtained; at least, this is a reasonable conclusion from their conduct. Perhaps live stock cannot be easily obtained hereabouts, but fish, vegetables & shell-fish can; & these are not brought off any more than the others. In the Japanese are to be seen the same curious mixture of politeness and unmeaning assent, half promising and non-performance, that is exhibited by the Chinese, and I think by all heathen people advanced to any degree of artificial society. The promise to perform & the excuse for not performing, are alike heartless; and can only be removed, I think, by a sense of fear. Probably it is indisposition to exert themselves which prompts this conduct, tho' too they may not be willing to tell us all the reasons and circumstances of the case.

Sunday, April 9th Off Kanagawa

The misty drizzle of the forenoon quite prevented all services on deck, but there was no work going on to speak of. The Commodore was taken up by a long discussion with Yenoske & Kenzhiro about going up the Bay. He says the President ordered him to go to Yedo, and he told the Commissioners a month ago that he was going, & they made no objection; other oft-repeated arguments were brought up, but no consent could be got out of him. They said Japanese laws were very strict, that great commotion would ensure, that the Bay was shallow, that the Treaty was signed, that the Emperor would be irritated, that as we had professed friendly feelings for them they wished us as friends not to go, and would regard it as a personal favor, and lastly, that very serious personal consequences might result, intimating almost jeopardy of honor & life if we thus implicated them. It was agreed that the ships would not anchor, unless they grounded; and then the whole party, as if willing to draw good from an evil, asked permission to go up in the steamers with the Commodore, that they might see the working of their machinery!

The fact is, the presence of such an argument, in the view of the officials, involves the intention & will to use it; for this they would do; consequently, mere curiosity to see Yedo cannot be motive enough in us to go, because it would not with them. The exposedness of their capital has startled them, and every subterfuge must be practiced to keep us from seeing any more than the surveying boats saw, for what motive can we have in such a nearer view than ultimate conquest or pillage or ransom? Judging us by themselves, our former forbearance, while possessed of so much power, can now be explained as having been exercised until the Treaty was signed; now we wish to learn modes of approach for future use, if we do not at present contemplate violence. Conscious weakness induces many a cunning fetch, which can only be explained by trying to place ourselves in the position of the weaker party; and the fear of ultimate designs is, I think, the leading motive of their strong objection to moving up. Yet after every dissuasive had been exhausted, it was not the less characteristic of them to ask a passage, not only to excuse themselves by the plea that they had done all they could to detain us, but to see what they had long desired to see in the working of the machinery.

Mond[ay], April 10th American Anchorage

By 8 o'clock this morning, the whole squadron were on the start, and bound for Yedo. The day was tolerably clear, and our Japanese visitors seemed to have little fear of any dreadful result of the day's excursion. By noon we had gone about ten miles from Yokohama, and seen the suburb of Shinagawa pretty distinctly, and its numerous rows or detachments of boats, not so many by far as I expected. The beacon we had so long had in view, proved to be a tower of temple inland and near Kawasaki, called Kawasaki Daishigawara, a place of resort and note. We went within about 8 miles of a long row of stakes stretching along in front of Yedo, but not so near as to prevent large junks lying inside of it, and turned about in 100 ft. water! If a man is Commdore, I suppose he can do as nobody else would, in order to show that he can do as he likes; and after all that had been said about going to Yedo, to say that we had left off four miles short of the surveying boats, & fully 8 of the city, was rather an imputation on common sense on our part. I was much disappointed; for, except a [~~few~~] lines of stakes and a long row of trees above Shinagawa & a smoky cloud above with plenty of junks & boats below, to indicate the probable position of the city, I saw nothing satisfactory. As one of the officers said, it should have been on the First of April instead of the Tenth, to make such a humbug appropriate. I have upheld and approved the Commodore's acts in most cases, where others have sharply ridiculed them, but this day's work was small enough. I have now been three times bound for Yedo, approaching nearer each time, & perhaps the fourth trial will land me there, or at least near enough to see it.

The "Lexington" drifted on shore when getting under way this morning, and the "Mississippi" returned to tow her off & bring her down to join the squadron. The scenes on board are said to have much amused the crew.

April 14th American Anchorage

No intercourse allowed with the shore here, and no visitors allowed to come near us, every native boat being kept away from the ships by a guard-boat, armed with authority to maintain non-intercourse. The "Macedonian" went to sea on the 11th early in the day, supposed to be bound for the Bonin Is. As she took some agricultural implements. The "Supply" & "Southampton" went down the Bay this morning, perhaps to Simoda. Surveying the anchorage goes on slowly by reason of the rough sea, and we are likely to be quarantined here a few days' longer. It is rather wearisome to be in sight of fields & headlands so long as we have been, & be debarred from seeing and rambling over them. Dr. Morrow stepped ashore a few paces yesterday from the surveying boats, & Mr. Gilsson reported it in such a way as to prevent even such an imperfect liberty, probably unwittingly on his part. He contrives, however, to do his part in annoying everybody below him, and does not keep his ship in any order at that; but it must be said that he has a hard school to manage. Mr. Draper & my brother [John P.Williams:1826—1857] were happily tranferred to the "Mississippi" two days ago, to serve as master's mates.

Went aboard of a junk lying off, stationed to guard us, in order to deliver a letter to Yenoske, informing him of the sailing of the two ships. We were kindly breceived, and shown whatever there was worth looking at, which was little enough. Nothing to eat, nowhere to sleep, and nothing to do, was about the whole of the matter for these sailors. They asked us when we were to leave, to which we replied they need not tarry any longer on our account. The main room had a thick deck and a pent-shaped roof; a dull fire was burning in a brazier or hearth in the middle of the deck, in a depression made for it, over which we found five or six of the crew crouching & smoking. The lockers, beams, & furniture of the cabin were all lacquered, and kept pretty clean too; but some of the lockers they opened for us were dirty. The capstan to hoist the big rudder was unshipped; it worked in two holes in beams, and was shaped like an oblong nine-pin. Six four-pronged grappling irons formed the ground tackle, and two well made coir hawsers; the tassel which hangs from the prow was, made of hair-cloth rolled around a mat. The tiller was larger in proportion even than the Chinese, and similarly hung; there were more points of resemblance to Chinese junks than I had supposed, and not a bit more of comfort.

Sat. April 15th American Anchorage

After telling us all that he had solemnly promised the Japanese officials that he would not go ashore at this anchorage, and refusing two or three applications for permission to stroll along the beach for flowers or shells, nor even to get beach-sand, the Commodore got tired of the monotony of this life, and took a stroll today on Webester

I, accompanied by five boats of officers and men, not to notice the two lopadogs. Of course this stroll, and the opportunity to stretch one's self, search for shells, pick up flowers, and dig dandelions for greens, was not the less pleasant because of these concomitants; the only objection was, it might just as well have been taken every day, at least on this little Natsu-sima, and all this fanfaronade saved. We collected several new & pretty plants and shells, and took a long row afterward to pass by the quarries which had been spoken of by the surveying boats, in which we rowed thro' much seaweed just like the Saragossa weed in the Atlantic. Coming back, we met Namura, with a message from "those high gentlemen" as Portman always calls the commissioners, hoping we would go to Simoda soon, as wood & water could be got there. Many charming views were seen in this row, and marks of great cultivation appeared on every hilldside; and amid so much that showed the goodness of God, we saw enough to prove that "only man is vile".

[Sunday], April 16th

Mr. Jones preached today, & for the first time since I left China I had opportunity to hear him; the text was "We preach to you Christ crucified", and tho' the language was sometimes high flown for the audience, the discourse was a plain, improving sensible exhortation. After the service, I saw one of the marines, named Goble, who wishes, if possible, to stay in Japan to learn the language, for the purpose of making known the testimony of Jesus to the inhabitants.

April 17th American Anchorage

Spent the forenoon in rambling over Natsu sima or Webster I. as we have named it. The position of this islet off Kanazawa [*sic*] facilitates its cultivation, and the most part of its surface is covered with thatching-grass, but whether this useful Arundo grows from wild stock, or is cultivated here could not be determined. Two fields of barley in ear and a patch of vegetables, are all the tilled spots. On the NW. corner is a shrine & a grave, both of them having rude statues; the latter covered by a shed, & having a bell hung near. They were both places of worship, & we met a fisherman & his family ascending the hill to offer their orisons. Some cash & bits of rags were laid near each of them, but no other offerings, nor any places of ashes as in China, from which I infer that incense sticks & paper are not as commonly burned. The suspension of rags around graves reminds one of the custom among Moslems, but no one which I have seen has the great number seen in the Mohammedan [welys].

We collected crustacea, fish, shells, insects, plants, everything which was worth carrying away, but the beautiful [detinia] were in too deep water to be easily procured; they looked very pretty, spreading their arms in every direction to collect their prey; and were so numerous as to give a gay appearance to the bottom. The low tide brought some dozen or two people to dig for clams, and the unblushing effrontery of these fishermen, as indeed of most whom we have seen, shows how much Japan needs the gospel of purity & love.

Tuesd[ay], April 18th Simoda Bay

The anchors were at the cat-heads before sunrise this morning, and the two steamers under weigh, coursing down this Bay of Yedo, probably for the last time. The day was smoky, so that we saw but comparatively little of the coasts, and were quite unable to discern any smoke from the summit of Oösima, which was wholly free from snow, & looked much less beautiful than when we passed it in February. A visit to this volcano would well repay the trouble. Approaching the eastern shores of the peninsula of Idzu, we sailed near enough to discern the villages along the sterile beach, but the background exhibited the industry bestowed upon it in the vast extent of terracing, which here far exceeded what we had hitherto seen. In one place, fifty steps of fields were counted, all covered with wheat. In the intervals, doubtless a large population is found to furnish hands to accomplish all this work.

About 3 o'clock, we came into the harbor of Simoda, and all were surprised at the variety of scenery and picturesque character of shores. The hills rise to a height of a thousand or 1500 ft., many lower ones, covered with trees, lining the beach, and adding a pleasing foreground in contrast with the barren & loftier mountains behind. The village of Simonda lies on the southwestern shores, and that of Kakisaki or [~~Oyster Pt.~~] Persimmon Pt. on the northern end of the harbor, both of them small towns compared with what we had been led to expect.

April 19th Bay of Simoda

The interpreter Tatsunoske came again this morning, but produced none of the things he promised yesterday;

in reality he is one of the most shiftless fellows we have to do with, and takes no trouble at all to get anything we ask for. In company with him came the prefect Kaheyoye and another officer named Nakadai Nobutaro, probably his spy. There were in all a dozen officials, all of whom as usual were glad to get a smack of toddy, wine & cake in the cabin, where they lingered a good while, talking and excusing themselves from doing or promising anything. The trip to Oösima was spoken of, but they had had no instructions respecting it, & could say nothing; the way is to go first & talk about the arrangements afterwards, so far as asking permission goes. In the afternoon, Mr. Portman & I went ashore to carry a list of provisions to be obtained, most of which Tatsunoske said could not be got, and a walk thro' the town confirmed his denial, so far as such a glance could prove anything. The town lies at the opening of two valleys, down one of which a small creek makes its way through the town, and forms by its mouth facilities for landing. At the landing-place is a small shrine under a large pine, and near it a hillside covered with trees invites [~~with~~] one to explore its grassy slopes. The town is regularly laid out at right angles, each street having a gate at each end, much slighter made than in China, and guarded with more care. The streets are wider than in Chinese towns, which makes the houses appear lower. The most of the shops and dwellings were of plaster, the roofs of tiling, and the fronts worked in raised white checker-work on a blue ground. The tiling is made of blue-black thick tiles, which lap over each other on the side, one edge being made doubly thick and umbo-shaped, so as to catch the thin edge of the next row; the ridges are therefore much smaller than in China, but more likely to leak, as the overlapping cannot be so well secured. A few houses were two-stories, but none presented indications of wealth, nor was there any place which seemed to be a market. The shops, so far as we could see, furnished a beggarly catalogue of sandals, groceries and such stuff; a total absence of the bustle of Hodangya proved the poverty of the port. The cancer of the social system was seen in the contempt shown to the women, but the power of the government was exhibited too in the sway excercised upon the crowds which thronged us. We went to the Riozhen-zhí 了 仙 寺 , a temple of the Budhists of considerable extent, having five priests in it, and many ancestral tablets; on these last are many names written, and most of them were varnished or gilded; perhaps they are orbate tablets. There was a graveyard near this establishment, and a small attempt at a garden, with a pond spanned by its tiny bridge, leading to the top of a huge boulder. The grounds and house were scrupulously clean, and the priest, named Nichizhi 日 净 or clean as the Sun, received us courteously. From this we went to four other tera [寺] or Budhistic temples, and one where a deified hero, called Goman-taro, of Yoritomo's time is worshiped. Votive offerings were hung around, and in a sort of porch were many pictures of shipwrecks, persons struggling with the waves, or just clambering ashore, and under them dozens of pig-tails strung along a board, the sacrifice of these rescued sailors, which they had cut from their heads to evince their gratitude. It recalled to mind the offering of Berenice, when Ptolemy was saved from shipwreck. Besides these, we saw a sword, a bow of large size, tablets & pictures, all given in as votive offerings, rendering the whole an interesting spot. The idols of the Japanese show more study and just idea of sculpture than the Chinese, so far as my observation goes, tho' neither are founded on just principles.

All the temples were situated back of the village

alongside of each other, approached by paved walks mostly lined with large trees. A row of magnificent [mowtans] proved that it was at home in Simoda; one flower was 10 in. across. A tree like a maple in its leaf, a purple magnolia, a spiræa, a plum, red & white Azalea, and a tree like the funeral cypress, were the principlal plants observed. All these establishments looked rather effete, as if they had once seen better days; and perhaps they were built when Simoda was the post of Yedo instead of Uraga, and maintained a large train of customs officials.

Thursd[ay], April 20th Simoda Bay

The storm has entirely prevented all visits, nor have any Japanese come near us; but the security of the habor has been well proven, at least for all north winds. The people of Simoda do not go out fishing [~~to~~] much [~~extent~~], & its shipping interests are plainly at a low figure.

In our walk yesterday[34], we were followed by most of the population, and all seemed healthy and well-fed. Ophthalmic complaints are prevalent, & small-pox has made its mark[;] the children are seldom pretty, & of the two, the boys are the most inviting; a few good looking girls hardly made amends for the scores of ugly or plain females —but a Houri or Hebe would never be able to stand roll-call after blackening her teeth & shaving her eyebrows. The women kept in the back of the crowd, as much from necessity as choice, I thought.

In one of the temples we saw 6 horses, that were haltered by a nose curb tied to each side of the stall; it held [the]m securely, as was proved by their [restiff] struggles at seeing us. Two gun carriages were also noticed here, apparently old and well taken care of. The insides of all these temples were varnished, nor have I yet seen a painted board or utensil in Japan. The priests we saw were cleanly dressed too, and one took pains to show us all over his domicile. Many prayer-books were observed in one of them, & their general furniture mostly resembled Budhist establishments in Canton.

April 21st Simoda Bay

The Commodore went ashore today with a small party to return the visit of the prefect, and was conducted to the 順法寺 Law-loving Monastery, until he could be informed of our arrival. The Japanese officials said we had come off earlier than had been mentioned, tho' they themselves had reached the Powhatan sometime before we left it, bringing a lot of provisions for us. At the temple we were received as civilly as the place afforded means, and when Kaheyoye appeared, he did all he could to entertain us, among other things sending out to let the people come into the precincts. About 500 or more persons came crowding around, fully one half of whom were women & girls, a few of whom were good looking. I do not think that Japanese features are as agreeable, when one sees hundreds of faces thus spread out before the gaze, as Chinese; the women's dress is not more graceful than the Chinese, & exposes the bosom more when the uncouth great girdle is loose, and the dress has been disordered. How many of these females were proper ones could not be known, but I rather think curiosity had drawn everybody out of doors to see us, and no restraint is put on their going out & in. Three or four of the better dressed, with their full proportion of girdle, more than a foot wide, & a knot behind that looked like a knapsack, and the hair done as neatly with a bow-knot flat on the top of the head, were brought into the room, & poured out a cup of saki for each. The discolored teeth of the oldest became more repulsive the nearer one could see them.

On leaving this place, we visited some other temples & walked around thro' several streets back to the boats, a large crowd of quiet spectators everywhere attending us, with the utmost order. One or two of the women most noticed at the temple, contrived to put themselves at several corners on our way in order to attract more attention.

April 22nd, 1854 Simoda Bay

General permission was given this morning to the officers of the squadron to go ashore, $2^1/_3$ mos. since arriving in these waters. A large number went ashore during the day, some of them taking long stretches and arduous ascents, which tired out the officials who were uselessly appointed to accompany us. The design of the Japanese authorities seems to be to watch us so carefully that no native shall supply us unauthorizedly with the least article, until the punishment of a few offenders shall deter all from violations of these restrictive prohibitions.

Dr. Morrow & I started for flowers and whatever else we could collect, taking the notherly valley from the town, and attended by four or five satellites, the chief one a well behaved man from Uraga, who had been at Yokohama, called Nakada Kadaiyu. The people thronged the streets as they did yesterday, but did not follow us. In one shop, we induced a woman to resume her weaving. She sat on a stool, & tying the woven end of the web around her body by means of a string passing from the end of the beam, she fastened one treddle to her foot, the other being secured to the floor; the loom was also made tight to the ceiling or the wall (I forget which) so that she should not pull it over. The shuttle was about fifteen inches long, sharp on one edge, and was used for both a shuttle to deliver the thread through, and a [slaie] to set it home, the thin edge being forced down upon the thread. The foot was drawn up under the other leg to alternate the threads, and make the web, which was of blue cotton, 15 ins. wide. The rudeness of this loom was doubtless owing to the poverty of the weaver, for better ones would be required to make the silks we have seen worn by officers.

A little further on, a blacksmith's shop presented a similar rude assortment of machinery; the anvil, forge, bellows & other things were so much like the Chinese as to excite surprise, for I should have thought some improvement would have been made. The men willingly showed us as much as we wished to see, and handled their tools like workmen.

Going out from the town, we reached rice-fields, now beginning to be turned over by the hoe, and took the chief road leading to the end of the valley. The waysides were covered with a carpet of little flowering plants, exhibiting a most beautiful variety of colors, and so abundant as to change the dun color of the ground as the eye glanced over it. A high stone gate-way on the top of a stone-walled plinth, formed the entrance to a temple on the top of the adjacent hill, nearly half a mile off & concealed in the woods; the hill was fancied to resemble Fusi, & the god was named Fusi Shengen Daibosats; if adoration paid at this gateway served the same as going to the top of the hill, the contrivance was a good one, [~~but~~] however much against our notions of architectural unity. The village of Hongoü, of a hundred houses, was beyond this gateway, where the headman came out to meet us & showed us his house, well built & having stone [~~walls~~] basement walls. Here wayside idols & pillars with Nammo Amida Bosats carved on them, showed, as they had, everywhere else, the prevalence of idolatry. A bowl containing young ferns in soak, called warabi [蕨], for food, stood near by the pillar. The pretty stream of Inedza-gawa ran through the village, the banks lined with shrubbery, and showing many marks of freshes here & there.

The walk was very pleasant, & we rapidly filled our book. The officer in attendance was sociable, and the people wer[e] not driven off; but to see one's fellowmen ordered about like dogs, their curiosity thwarted and convenience disregarded as if no more consequence than a spaniel's, humbles the race in one's own eyes, and imparts a feeling of reproach as belonging to the same race from merely beholding this outrage on the dignity of man. A people that will tamely submit to it must have been schooled a long time by their rulers, & given up at last in despair.

The path leads along the river bank, at the base of the hill, & two undershot water-wheels, each turning four pestles to hull the rice, showed that the people deemed its water useful for something else besides irrigation; they were placed quite out of the reach of the freshes, & fed by two races. In this village, some fine [~~stone-~~] walls of dressed oblong stone showed that such work is well done where the material is handy; one large stone house was going up, the building precincts being surrounded by a high wattle of bamboo to keep of intruders. A fire-proof granary and a garden-wall of the same material in another hamlet gave evidence of some wealth, but [~~the~~] most of the peasantry in this valley seem to have few extra kobangs for such purposes. Pine trees trained in a fanciful manner in stories were found in a few yards, one of which had been under tutelage for fifty years, and made very little progress all that time, while its untrammeled congeners congeners of half that age on the hillside were huge trees towering far above all others; the contrast made one think of the effects of similar misdirected toil upon the human mind or body deforming and vitiating what God has made upright and complete.

In this village, I noticed a Trimurti image drawn on a piece of paper on a door as a charm; it was called 庚申神, and was Budhistic & not Brahminic. I have never seen one like it before. Charms of fern tied with rice-straw to a wisp stretched over the gate still remained from new year's time, and the names of gods written on paper pasted at doorways, besides pillars, terminalia, & idols, met one constantly, all proving the hold idolatry has upon the people.

We now left the valley, and ascended the hill towards the sea, finding new vegetations & several flowers as we ascended above the plain. From the top, the beauty of this valley & Hongoü appeared in all its glory, showing how God has given its inhabitants rain & sun, seedtime & harvest, all in their season, filling their hearts with food & gladness, if haply they might seek after him. Alas, how sad is their ignorance and degradation! A pleasant air greeted us on the elevated paths, & erelong, we reached Shirahama 白濱 a village of 150 houses, lying on the beach seaward. Most of the inhabitants came out to see us, as we did not enter it, and enjoyed their first sight of foreigners with evident satisfaction; they were a rough set, sunburnt, dirty, brawny, and half clad.

The path now led up the hill towards Suzaki, a town of 230 houses near the beach & the entrance to the harbor, but we turned off by a road leading to Kaki-saki 柿崎 or Persimmon Pt., in the harbor. The hamlet of Soto-ura or Outer Hamlet, of 50 houses, lies near Suzaki. These upland hillsides were under a high degree of cultivation, mostly wheat & barley fields, and presented as charming a landscape as the eye usually rests on — sea & land, valley & hill, woodside & heather, grain, vegetables and shrubbery, hamlets, farmsteads, & town, all alternating in the most picturesque manner. We met many groups of men, women & children returning from Simoda, & it was really distressing to see the manner in which they crouched & skulked out of our way at the behests of our keepers. At a wayside teahouse where they asked us to stop, for they were well tired, I was allowed to buy cakes and candy, and had a pleasant chat with the people, telling them how I had entertained & fed their shipwrecked countrymen in my own house. Even here, a man was stationed a rod or two ahead to order the passers by go aside, which they did, much as they wished to see who was there. — We got back to the ship before sunset, weary & gratified with our walk.

[Sunday], April 23rd

While service was conducting on board the Mississippi today, Com. Perry took the trouble to send his flag-lieut. over to the ship to tell Capt. Lee that the ensign was lower at the peak than the church-pennant, & take him to task for the misdemeanor. He had hardly gone, and Mr. Jones had just begun his sermon, when signal was made to furl sail, they being a-drying, which of course broke up the whole service. If God is long-suffering to the heathen in bearing long with them, how much more so is he to those who know his law & disregard it! Soon after this, Perry was off ashore in his barge, carrying guns to get birds with, totally indifferent to the regulations of the service, let alone any higher law. The

Southampton is a-painting today too.

[A name card written in Japanese, of the above-referred temple 了仙寺]

Mond[ay], April 24th Simoda Bay

Mr. Bent went ashore betimes this morning to see the prefect. One of the objects of the visit was to advise him to issue orders that the officers of the squadron be not followed by Japanese officials in their rambles thro' the country, nor have the people shut their houses up and run when foreigners came in sight. He said that he had left Yokohama so long before the treaty was signed, he was unaware of the views of the commissioners respecting the attendance of officials when the Americans went into the villages, and must report for instructions; at Yokohama it had been done, but he, himself, having much confidence in us, was inclined to try how it would work, for it was a serious burden on the officials under him to accompany us here and there. As to people running or staying, when they saw us, it was a matter he could not control altogether, but he would issue commands to let them know they had nothing to fear from the foreigners, but were to report any misdemeanors. Another point was to procure a junk for the accommodation of a party to visit Oho-sima, and examine the volcano; and also to get three or four rooms in town for the convenience of the officers. The rooms in the temple of 順法寺 at the back of the town were visited after the interview was over, and made an offer of by the priests, somewhat to their inconvenience I guess, judging by their looks. The disposition to accede to our requests on the part of these officers augurs well to pleasant intercourse. They seem to be more particular respecting trade than anything else.

After this, a party of us started to follow up the valley south of the town, and took a course along the beach for a while, and then struck across the hills till we reached a place called Nabeta, in a secluded dell, where not much of this world's riches or ambition have yet come, and the inmates of its seven houses proved that no great amount of its cotton fabrics had reached them. The valley was soon to be turned into a huge rice field, and one man was ploughing with a simple plough, made of a beam with a crooked handle to sustain & guide a spare shaped like a big spoon, which turned over the earth five inches with much effect. It was not so effective, however, as the deep hoeing of two or three lads with the ploughman, whose three-prolonged dung-rakes turned over the wet land very easily.

A short walk carried us to Okagawa, a hamlet of about 60 houses, beyond which was a large temple inclosed in almost a complete solitude, where contemplative Budhists might drone away their lives in total listlessness. It was called Shio-riu zan 小走山, & the temple Soö yin. Budhism must have a deep hold upon the minds of the Japanese to induce them to erect such structures in wilds like this, so far from the abodes of men. The five priests living here keep the houses & grounds in clean condition; one of their acolytes was only four or five years old. Near the place, as we left it, a waywise god called Doöso-jin, attracted our notice from his holding a sceptre in his crossed hand, and his head being covered with a sort of crown, from which a broad cape descended to cover his back & shoulders. Many of these [terminalia] offer curious subjects of speculation.

From this, a rugged mountain path led us over to the valley north of Simoda, to the village of Hongo, where we met many officers walking, and joined them. This valley is incontestably the most beautiful in this vicinity.

April 25th Simoda Bay

Two Japanese came aboard last night to get a passage to the United States in our ships, but the Commander declined to receive them, unless they had previous permission from their own rulers to do so. They had previously sent a well written letter intimating their desire to go, & willingness to do anything on board. This letter was to the following purport:

"Two scholars of Yedo in Japan, named Isagi Koöda市木公太 [Yoshida shouin 吉田松陰][35] & Kwanouchi Manji 瓜中萬二 , present this letter to the high officers and others who manage affairs. That which we have received is meagre and trifling, as our persons are insignificant, so that we are ashamed to come before distinguished persons; we are ignorant of arms and [~~their~~] uses [~~of~~] in battle, nor do we know the rules of strategy and discipline; we have, indeed, uselessly whiled away our months & years, and know nothing. We have heard a little of the customs & knowledge of the Europeans & Americans, & have desired to travel about in the five great continents, but the maritime probibitions of our country are exceedingly strict, and for foreigners to enter the "inner land" or for natives to go to other countries, are alike among the immutable regulations. Therefore our desire to travel has been checked, and could only go to and fro in our breasts, unable to be uttered, and our feet hampered so as not to stir. This had been the case for years, when happily the arrival of so many of your ships anchoring in our waters now for many days, and our careful and continuous examination of the kind and humane conduct of your officers and their love of others, has excited the desire of years, which now struggles for its exit. We have decided on a plan, which is very privately to request you to take us aboard of your ships and secretly carry us to sea, that we may travel over the five continents, even if it is disregarding our laws. We hope you will not regard our humble request with disgust, but will enable us to carry it out; whatever we are able to do to serve you will be considered as orders as soon as we hear it. When a lame man sees another walking, or a pedestrian sees another riding, would he not be glad to be in his place? How much more now, [~~when~~] since, for our whole lives, we could not go beyond 30 deg. E. & W., and 25° from N. to S., when we behold you come riding on the high winds & careering over the vast waves, with lightning speed coasting along the five continents, [~~do~~] [~~we~~] does it appear as if [~~now~~] the lame had a way to walk, or the walker an opportunity to ride! We hope you who manage this business will condescend to regard and grant our request; but as the restrictions of our country are [~~severe~~] not yet removed, if this matter becomes known, we shall have no place to flee, and doubtless [~~well~~] must suffer the extremest penalty; and this would greatly grieve your kindness and benevolence of heart to your fellowmen. We trust to have our request granted, and also that you will secrete us until you sail, so as to avoid all risk of danger to life; and when we return here at a future day, we are sure that what has passed will not be very closely investigated. Though rude and unpracticed in speech, our desires are earnest, and we hope that you will regard us in compassion, nor doubt or oppose our request. April 11th."

Inclosed was this note: "The inclosed letter contains the earnest request we have had for many days, and which we tried many plans to get off to you at Yokohama in a fishing-boat by night, but the cruisers were too thick, & none others were allowed to come alongside, so that we were in great uncertainty what to do. Learning that the ships were coming here we have come [~~here~~] to wait, intending to seize a punt to come to the ship, but have not succeeded. Trusting that your honors will consent, to-morrow night after people are quiet we will be Kakizaki in a punt at a place where there are no houses near the beach; we sincerely wish to have ou come to the spot to meet us, and thus carry out our hopes to their fruition." April 25th[36].

They came up the ladder by the help of the quartermaster, but unluckily their punt slipped away as they left it and drifted off. The commodore was told their errand & about the above letter, but he could not take them without violating the spirit of the treaty. It was a severe disappointment to them, but I told them that other ships would come here in which they might get off, and that they must not be oversorry at this refusal. They were put ashore in a boat, & directly to leeward in order if possible to get their own, but it was too dark to see it. They were more ordinary looking men than I had expected to see, but evidently men of education, 23 or 25 years old, neither parents nor children to keep them in Japan, and were probably just what they said they were, eagerly wishing to go to U.S., tho' some said they were thieves, others spies sent by the officers to see how far we would keep the treaty, and others that they were refugees from justice. I am afraid the loss of the punt containing their swords, &c., will involve them in trouble; it was picked up by one of the cruisers in the habor, & some officials came aboard to inquire about it, but of course we told them nothing. (See May 7th.)

Mr. Pegram & Mr. Jones went with us today to Suzaki,

a town of 230 houses, situated near the beach just outside of the harbor, and offering nothing of interest; much stone & firewood were lying along the beach to export, some of the former glistening in the sun from the quantity of crystals of pyrites in it. A short walk, during which Mr. Jones returned on board, brought us to Soto-ura, a miserable hamlet of 30 or 40 houses, the inmates of which received us pleasantly. Beyond this was a quarry where large blocks of bluish amygdaloid were slid down the hills in a tram-road; this rock would make a fine article of export to California for building. Going on to Shirahama, we tarried awhile in a temple, the walls of which were nearly covered with paintings of various sorts, all labeled 奉納 , and showing the low state of the fine arts of nothing else. The officials were, apparently, glad of a chance to do something in this village, for they made many efforts to keep everybody away from us, and accompanied us over the hills to Hongo. The walk was a pleasant one, & afforded more chances for picking up new words than new flowers. It was amusing to see the women skulking under the banks to get out of our way, and still desirous of getting a look at the dreadful men.

April 26th Simoda Bay

At last the printing-press is to be set agoing, and I have been busy with it today.[37] In the afternoon I went among the graveyards attached to the temples in Simoda, to learn something respecting the inscriptions, taking Lo with me.

There are ten Budhist temples in this small town, and they all are in good order, served by many priests, some say nearly 50, others half that number. The largest is called Hozhun san 法順山 in it[;] the Commodore has a room or two; the others are named Dai-an zhi 大安寺 or Great Peace

Monastery; Hongaku zhi 本覺寺 [;] Hoöfuku zhi 寶福寺 or Precious Peace Mon.y; Kaizhen zhi 海善寺 Ocean of Goodness M.; Toöden zhi 稻田寺 Rice-field M.; Fukuzhen zhi 福泉寺 Fountain of Happiness M.; Chioraku zhi 長樂寺 Long Joy M.; Rigen zhi 理源寺 Source of Reason M.; & Chiome zhi 長命寺 Long Life M. Besides these, the Sintu mia of Hachiman 八幡宮 dedicated to an ancient hero. All the monasteries have burying-grounds near them, containing a variety of monuments as with us. Most of them are old, and the inscriptions in the trachyte covered with lichens and illegible. In one place, a family of rank and all their servants were mouldering in the same inclosure; the tombs of the chiefs were cut in the shape of a chess-king, a succession of cubes and spheres one over the other, surmounted with a cone. Many persons are buried under each monument, their bones being put into jars; otherwise they are buried in a sitting posture. At the entrance of the 法順寺 is a new monument, with an invocation to Budha in gilt characters, containing the names of several persons, and the following stanzas: [4 lines of Chinese characters with the following English translations:]

What permanency is there to the glory [~~and~~] of the world?

It goes from the sight like hoarfrost before the sun.

If man wishes to enter the joys of heavenly light.

Let him smell a little of the fragrance of Budha's canons.

Over some of the epitaphs was to be seen the coats of arms of the deceased, but generally, the words 妙法 Wonderful Budha! commenced every gravestone; these stones are cut out of the awo ishi, or greenstone, and soon become covered with lichens. At the entrance of one of the monasteries, a pillar was set up, giving notice that 不許葷酒山門 "all meats & spirits are disallowed to enter these gates." Slips of board stood near some of the graves, containing inscriptions; one was, "[~~If~~] [~~you~~] [~~wish~~] Whoever wishes to have [~~your~~] their merit reach even to the abodes of the demons, let them with us & all living become perfect in the doctrine", by contributing money. Another was, "The wise will make our halls illustrious and the monuments endure for long ages", by giving money. A third is, "Whoever enters the paths of Budha will doubtless enjoy his fill of pleasure." 圣云是人於佛道足樂無有疑也。 These hints received more direct encouragement in the boxes suspended inside of the temple, which were labeled "for feeding hungry demons", and a promise attached in one instance that "then their merit would be considered". Images of Budha were placed near the graveyards; one well carved was named Illimitalbe Glory Budha 不斷光佛 & had a post behind it declaring, "that whoever wished to thoroughly learn the three ages and Budha, must examine the restrictions with his whole heart". Some of the epitaphs notified that the deceased had recited 1000, 2000, or 3000 volumes of canonical books, intimating thereby his title to felicity.

The mia [宫] dedicated to Hachiman had no priests in it, and was differently arranged. Two weather-worn stone guardians of fierce aspect defended the entrance at the bottom of a long avenue who looked more fiercely from the lichens on their faces, a belfry stood on the right, the bell being struck with a long heavy stick hung so as swing against it. A porch contained the votive offerings already mentioned, led to the temple, [~~in~~] which were other things, as sword, bows, fan, mask, pagoda made of cash mailed on a board, also offered with the name of the donor. A subscription paper, about 20 ft. long, containing names of donors, proved the universality of the taxation of the people for idolatrous objects.

April 27th Simoda Bay

Dr. Morrow & I set out this morning to follow the Inodza gawa to its source, and take fish in its waters. We saw nothing new, except a shop where a paper-maker was preparing the bark of the Broussonetia by soaking it, and laying it on bamboo molds. At the village of Nakanozhi, or Hongo, of 50 houses, we bought some [~~peas~~] red & white beans & raw hemp, the headmna standing by to assent thereto. The fish would not bite; only a few gobies were caught, and the lines put up. At Mits-tskuri, a hamlet of 20 houses, the river divided into two-runlets, whose course led up two winding valleys, as far as we could see, the sides of each being terraced to a greater extent than I ever saw elsewhere. It was a charming sight, for such a long vista of cultivated fields, now ripening to the harvest, in this manner were spread before the eye more beautifully than the same extent in a single field would have been. The background of forest exhibited great diversity of tints in its foliage, adding a new style of ornament to the landscape. The people received us cordially, and no officers prevented our intercourse; their houses were scattered over the country, as if there was nothing to fear in any degree of isolation. We took a short cut back, accompanied by two pedlers, and had some sport with them; as we were trading off a handkerchief for 5 eggs, a ragged laborer came by, and I was most humbly besought by one of them to give him two cash as a bribe to silence him in the village, and it was amusing to hear the injunctions given & promises made by the parties. On the hill-top, this & 3 other h[an]dk[erchie]fs. were carefully wrapped inside of their stockings before entering the town. At its foot, a small village lay, at which our friends had marshaled the entire population, & thus the people of Rendai zhi 蓮臺寺 or Lotus-terrace monastery had a full view of the foreigners, when they came by.

April 28th

The rate at which it is arranged that our coin shall be taken renders the price of all articles which are procured [~~nearly~~] more than double what they are usually sold at. The Commodore has agreed upon 1200 cash as the worth of a dollar, while the people pay nearly 3000 as the equivalent of the same weight of silver; when, therefore, we buy articles in the shops for silver, the people reckon the money at their valuation of cash, but when, according to the arrangement, the shopman takes them to the guardhouse where the official comprador has his office, the other rate of exchange is adopted, and cheating is supposed to be extensively carried on, while in fact much of it is owing to this unequal valuation of the dollar. The complaints of the men who buy with silver are loud, while those who take cash ashore [have] no complaints to make, because they have already their discount on board. Articles of fine workmanship are few here, but the best are rapidly sold, and if the officials only took a moderate percentage on them, I would let the shopmen have the profit of the exchange for a while, until a juster valuation was made out.

Sat[urday], April 29th Simoda Bay

Dr. Morrow made a small collection of algoe and

soft corals today along the beach; the existence of the latter in this latitude shows the warmth of the seas. An excursion was made by the surveyors to the rocky islets off the mouth of the harbors, which were found to be much greater protection to it from a south & SE swell than had been supposed. This harbor is now nearly finished, and the two sunken rocks in it are to be signalized by buoys, in doing which the Japanese claim the duty of bearing a part of the expense. This proposition on their part shows a higher sense of care and protection for shipping than we had given them credit for, & more than the Chinese have ever exhibited; and the same desire to improve will soon do away with the restrictions which now impede the natural extension of their commerce. How much, too, can be hoped from the introduction of true knowledge and religion, which I can hardly doubt are in some way to be brought among them.

In going about among the shops, I found that the household was almost always under the same roof; and the female part of it had something to say respecting traffic. In most shops, the goods are kept in drawers when they can be, and only coarse crockery, grain, bamboo-ware, and other coarse, cheap articles were seen. There is no counter, but the two parties sit on the same mat to trade, and only a few precautions are apparently taken against theft. No money-changers line the streets as in Canton, nor do we see anything hawked thro' the streets; a few peddlers are met in our rambles, but there is much less of such trade than in China, and not nearly such an air of industry & bustle. Only one school has been found, and the boys who throng around us are seldom able to read, so far as I can ascertain. In respect of slovenly habits, they and Chinese lads of the same class are about the same; while both sexes, old and young, are if anything, more degraded in respect of morals; the dress of this people is far more excetionable and less modest than the Chinese.

Sund[ay], April 30th

In consequence of a threatening squall, there was no service on board this or the other ships; and the day was quietly spent. Large numbers of officers & men went to Simoda, where trading was briskly carried on, as if the obligations of a Sabbath had no stringency here, and there was no need of observing it. If officers even set an example of regard to a Sabbath, the effect would be better than the promulgation of any law.

Mond[ay] May 1st Simoda Bay

Before going out of town, we went into a few shops, in most of which we found nothing worth buying or hardly worth looking at. The common ware is very much like that used by the Chinese, nor do the shapes of the dishes differ very much from those seen in China, the same customs inducing the same forms. The common lacquered cups & trays used at meals present nothing unusual in style or excellent in workmanship. In one shop a good-natured pair showed us most of their wares, sold us a lot of raw hemp & a box and exchanged 8 large cash for the same worth of Chinese coin. Her tidy daughter was standing by, a good comment on her housewifery, to whom I gave a picture-book I had, much to the joy of the mother. We saw nothing worth notice until we had reached the village of Nakanoshe, where we wished to inquire what direction to take for the paper-tree but no one would show us or go with us. However, we got a sight of a rice-hulling mill, and were talking with the owner, when a woman came running up, & began to tell me she was the shopkeeper's wife, of whom we had bought the flax and box, and had been ordered by the officers to get back the large cash she let me have, & give up the picture-book too. She had run a mile to overtake us, and begged me to let her go back to Simoda with the 8 cash, holding up the string I had given her; of course she could not be refused, but this exhibition of littleness on the part of the Japanese officials shows the character of their espionage and oppression. I could not learn why she told them of the transaction at all.

Going up over the hills beyond the village of Rendai zhi, the boys showed us the paper-tree, just in time to find flowers & fruit on it. We saw the tools for beating the pulps out & jars for holding it, and molds, with a heap of the fresh paper, some wet & others drying on boards in the sun.

The people received us kindly in all the hamlets far up the valley over the hill, and we stopped at a bridge where the Inodza River was about a good leap across, being pleasantly told by a pedler that there was nothing at all beyond, not the least thing, & the day was far spent. It was one of the best rambles I have had, the people accompanying us along the path, asking for cash, for autographs, or for information, in the most courteous manner.

May 2nd

The "Macedonian" came in from the Bonin Is. This evening, having been three days in returning from Port Lloyd, where she left an anchor and bought three-score turtles, all there were to be had. The population has decreased since last year, & the U.S. Consul left there [~~last~~]

[~~year~~] by Perry has gone off in search of better quarters, making thereby a good move. All the stock left there has disappeared, nor have the seeds come to maturity; and except the turtle & potatoes brought away, there is little new or interesting added to our present stock of information or stores.

In a small cove near the village of Oöura, where we went to collect seaweed, we found trap-rock in regular pentagonal basalt, the columns divided from each other distinctly, dipping about 80° S. The rock was not coarse-grained, nor was it very hard, many perforations being seen in the base made by shell-fish. In another place, the trap was very coarse, full of cells, and resembling scoria.

Wed[nesday], May 3rd Simoda Bay

Our walk today led us by a small plat of the Bignonia tomentosa or kiri of the Japanese which they cultivate for its oil to use [as] varnishes, mixing it with the juice of the varnish tree. Several patches of the tree have been seen at various times in our rambles. Near the town, we passed a small shrine or mia dedicated to Shio-ici-imari-dai-mio-jin, which possessed nothing of interest, except two doorway guardians of foxes, carved in a passable manner, with very bushy tails. The shrine itself is on the hill-top, reached by a flight of stairs, and as usual surrounded by trees, recalling to mind the idolatry of the old Jews among their groves and under every green trees. Why these demigods are enshrined in hilltops is not very clear, unless the people choose pleasant places for themselves in worshipping dumb images. The next thing of interest today was a visit to a school-house, in a temple, where 14 low writing tables were spread about the room, high enough to write on when sitting on the floor. The boys come at 8 o'clock, & go home at 11 o'clock; the afternoon session is from 12 a.m. to 4 p.m. In all about 50 boys come, & the teacher receives presents from his pupils as they please. In the room stood a gigantic image of Budha, with the past & future Budhas at each side; all were of copper, the larger about 20 ft. high, the others nearly man's size, all in a sitting posture. No priests lived here. There are five other schools in Simonda for boys, where they learn writing, keeping of accounts, and different styles of epistolary composition. Two others have been met out of town.

Going southward along the beach, a new style of volcanic rock was seen in the nodules of trachyte & trap imbedded in a conglomerate of scoria & sand, some parts showing little trap & others little sand; sulphur was evidently present too, proved by the efforescence. The cliffs along the shore in these places presented a black color; many large masses were strewed along the beach, while in the sandy portions, the people had cut out rooms, in which they stored seaweed, or found a refuge themselves. Some of these lodges were a hundred ft. up the cliffs, & it was not clear how they were reached. Many persons were seen drying seaweed on the sand — a repulsive sort of food eaten by equally coarse people. Near the town of Kichisami, a scattered hamlet of 100 houses, we were overtaken by the rain, and at a hovel we went into, amused ourselves & the poor people by talking to them, writing, and giving them bits of woolens. By the time we reached Simoda, we were thoroughly soaked through, and glad to get off to the ship. The peasantry have everywhere received us with the utmost kindness in all our rambles, but today they took some pains to enterain us.

May 4th

The "Lexingon" left the habor this day for Lewchew, where she is to remain until the rest of the fleet joins her. It rained most of the morning, and after dinner I went ashore to see the prefect respecting letters of introduction to be sent to Hakodade by the three ships, which the Commodore sends ahead of the flag-ship. The necessity of doing all that the time allowed in supplying the bazaar was also urged, and of taking off the restrictions which impede the free intercourse, to which he gave partial assent. The power exercised over the mass of people by their officers must require a large force to uphold it, or else the fear produced by this system of espionage renders each individual so isolated, and conscious that he has no alternative but entire submission, that the police is less than would be necessary in any other country. What requires a powerful army in Austria is done in Japan by rendering every person isolated, and thus accessible by a single order backed by only the messenger who takes it. Yet the introduction of free opinion here would soon show the rulers the need of changing their policy, and perhaps a revolution would gradually be made by the diffusion of such sentiments among all classes without a convulsion. It is an encouraging view to take of the matter, that God is able to overrule all these adverse powers so as to bring about the extension of his truth and thereby the best weal of these eastern nations.

May 5th

The rain of yesterday left the roads very muddy, and the people of Shira-hama were if anything dirtier than ever, for here as everywhere else poverty and ignorance are accompanied by dirt and vice, and whatever cleanliness may be seen in the houses of the better sort does not pervade the laboring classes. However, they treated us with more kindness than last time, and the officials kept away entirely. Our walk led us to the top of the highest hill in this region from whence the peculiar ridged formation of the country-large ranges forming watersheds, the sides of which were intersected by ravines, mostly well wooded, while the main backbone was bare. On this peak grew a grove of large pines, and superstition had here erected a shrine to Zhi-zo bosats, a Budha, whose image stood in solitude & gloom; the room was covered, almost with pictures and inscriptions, besides a good many cues, offered perhaps by sailors and others, whose gratitude had led them to take the trouble to mount Taka-san there to hang them up.

May 6th Simoda Bay

One of the men fell from the foretopsail about noon yesterday, and was dreadfully bruised that he died about sunset, having his reason to the last, for in his fall his head was untouched. This morning, the officials came off to inquire respecting the casualty, and our wishes in respect to the funeral & burial. I & Mr. Portman accompanied them ashore after breakfast, and they stated the matter to the prefect, who said that at present only temporary arrangements could be made for a burial-ground, and he must await the arrival of the commissioners before definitively setting apart a spot of ground for a foreign cemetery. He and the others decided on burying the body at Kakizaki, and a place was cleared in the cemetery attached to the Yoku-zhen-zhi 玉泉寺 in that village, and the funeral took place about 5 p.m., the whole population being present to see the ceremonies. At the same time, the Commodore was entertaining the townsfolk at Simoda with the band. The tombs and inscriptions in this cemetery were different in many respects, from those at Simoda, tho' on the whole alike. On many of the epitaphs, the phrases, "returned to the original", 歸元 "returned" or "joined to the company", 同會 "gathered to the original", 皈元 "annihilated" or "absorbed", 空 [blank space] were inscribed instead of the exclamation "Wonderful Budha!" The words 信士 "believing scholars" & 信女 "believing woman", were joined to the name, followed by the word 座 "seat"; some epitaphs had a space left for the wife's name to be added; and many gave the names of the children as well as parents, all on one face of the stone. The monument for priests resembled sugar-loaves, and were mostly covered with lichens; others were surmounted with a carved stone to represent a roof or a hat, and all had a little trough to hold water cut in the plinth, [~~below~~] before the epitaph. Small images called chi-zō 地藏 were placed here and there, which I understood were worshiped by the dead; it was a Budha in the attitude of worship. Over some, a mowtan or a lotus was carved, in a few cases a coat of arms, and the swastica as often as anything else. The stones were crowded close to each other on the sides of the hill, so as hardly to allow passages between the rows, and the path was much blocked up by flower-jars, boards, and other rubbish. The groves of pines on the hill renders the spot a sheltered one, and it is a more desirable place than one in Simoda. Here, surrounded by Japanese, lies the body of poor Parish, who had ran away from his parents in Hebron. Connt., and had given them no notice of his course since — an instructive commentary on the rashness of disobedience to parents.

May 7th

Rev. Mr. Jones gave a discourse on the casualty and warning Providence just fresh in our minds, which was calculated to do good, & I hope will be blessed to some of the man's mates. The Commodore heard last evening that the two men who had come off to the ship on the 25th ult. were caged on shore [Yoshida Shouin 吉田松陰 :1830—1859, and Kaneko Sigenosuke 金子重之輔 :1831—1855]. Mr. Bent & I went to see them this morning, but were too late, as they had been taken off to Yedo at daylight. The keeper of the house told us they were imprisoned for going aboard our vessels, and had been detained here until orders were received from Yedo; but I learned nothing as to the probable punishment they are likely to receive, tho' I fear the worst. The cage was about 6 ft. long by 3 wide & 4½ high, quite large enough to sit & sleep in, and entered by crawling thro' a low door; it is probably just such a cage as McCoy & his fellows were at last shut up in. It seems that criminals are not examined in Simoda, but are sent to Niraiyama, a town about 20 ri north, where Taro zaiyemon, the deputy of Toda, the prince of Idzu, lives, & judges them. The present case, however, did not come under his jurisdiction.

Mond[ay] May 8th [Simoda Bay]

Mr. Pegram accompanied us up the valley beyond Rendai zhi, in which we met many well disposed people & some new plants.

The farmers are preparing their rice plats for sowing the grain, & laid a layer of dock & other soft leaves on the water mesh, which was so soft that it was easily trodden beneath the surface by a man walking over it with a pair of snow-shoes, which he held upon his feet by means of a string passing round the forepart, his feet bearing down on the back. This subsoil would soon be decomposed, and furnish support to the growing shoots whose roots would thus be more easily lifted. It was a singular operation to see the naked fellows dabbling about in the mud, & preparing these plats. Many sick persons applied to us today for relief, and we could only ask them to come aboard ship, which I am afraid none of them will be allowed to do. One was a case of bronchitis, one of rheumatism, & several ophthalmic cases as well as other diseases of a minor kind. I told the people I thought many of their ailments of the eye were ascribable to the custom of shaving eyebrows of the women, & that to keep them clean would be one way of curing them. A physician would find a large field for his efforts among the Japanese, but I doubt his being allowed to practice.

A pictorial representation of our squadron & description annexed, and an account of the war between England & China, were seen today by officers, but neither of them could be purchased; the authorities are so whimsical in their conduct that it is impossible to follow them, or account for the orders by the actions of the people. A lot of ginseng was bargained for at 160 cash, but Tatsnoske would not let it go for $3.40.

Tuesd[ay] May 9th Simoda Bay

Three of the officers went on a hunting excursion yesterday, and managed to kill one live pheasant, shoot off a talk feather from another & buy a dead fox, for which they had a tramp of 25 or 30 miles. Getting back to Simoda about 9 p.m., they concluded to remain ashore all night, but tho' the priests were willing enough, the officials and interpreter were not, and came in to order the party to go off to the ship, using violent language and behaving in a most impertinent manner, besides calling in a guard of soldiers and having lamps placed in the room. Every effort was made to appease them, but unsuccessfully until one of the sailors was told to get out the arms, whereupon the men & the lantern-bearers soon vanished, their superiors having gone before. The guard remained outside of the room all night, and at dawn the officers came aboard. The whole transaction was so impudent, that notice was taken of as soon as it was convenient, and Mr. Bent, with Nicholson & Tansell were sent in uniform to report to the prefect. Tatsnoske was half drunk last night, and it required some sharpness to make him speak out to his superior, who at first wished to shift the matter to the commissioners, and to inculpate us as also having done wrong in stopping ashore, as having violated the treaty, and also as having mistaken their design in placing a guard. However, the might being on our side the right was too, and by a threat of going to Yedo an apology was drawn from the prefect, with a promise that such usage would not be repeated, & the officers might stop ashore whenever they pleased all night. In truth, the insolence of the officials was the principal point to be checked, and they will soon learn we are not going to be treated slightingly with impunity.

Of all heathen nations I have ever heard described, I think this is the most lewd. Modesty, judging from what we see, might be said to be unknown, for the women make no attempt to hide the bosom, and every step shows the leg above the knee; while the men generally go with the merest bit of rag, and that not always carefully put on. Naked men & women have both been seen in the streets, and uniformly resort to the same bath-house, regardless of all decency. Lewd motions, pictures, and talk seem to be the common expression of the viler acts & thoughts of the people, & this to such a degree as to disgust every body. Alas for the condition & excellence of a simple, heathen people, dreamed of by moralists who never saw what they prate of. Truly, the long-suffering of God is infinite, or Japan would spew out its inhabitants. I hope there is good still for them in store.

May 11th Simoda Bay

Yesterday a driving storm of rain kept everybody on board ship, & not till this afternoon could we easily go ashore. I saw today a board obtained from the two imprisoned men, which seems to have been written for our inspection, tho' the language is guarded, and would be almost inexplicable without knowing the circumstances.

"When a hero fails in his designs, his conduct is then regarded like that of a thief or outlaw; (we have been) seized publicly and then guarded, darkly imprisoned (here) for many days, treated roughly and proudly by the village elder & headmen, whose harshness is very great. Yet we can look up without reproaching ourselves; and it can now be seen whether a hero will act like one.

"Since a journey thro' the 60 countries (Japan) was not enough to satisfy our desires, [~~we~~] [~~wished~~] to travel in the 5 great continents was once our hearts' desire; but suddenly we missed our aim and are now fallen into a half-sized house, where eating, slepping, resting, sitting, are all difficult, & escape impossible. If we weep we appear like fools, if we smile we are deemed to be rogues. Alas, silent we must rest!"

No clue will probably be obtained to their fate while we are here.

Frid[ay], May 12th

Everybody who could leave the ships went ashore this morning to buy or settle for things they wished, and to take a final walk thro' the town. I went a way thro' the valley with Mr. Pegram and Lanier, and enjoyed the ramble as well as found some new plants. On board, many hundreds of dollars were paid to the officials for the supplies furnished, which at the high rate of exchange, left them an enormous profit, as much as 300 [pr.ct.] in some cases. This unfair mode of trade doubtless will henceforth be changed by making our coins worth more cash.

Wed[nesday] May 17th Hokodade Bay

The two steamers left Simoda early on the 13th, and had a very pleasant trip to this port. A shoal, supposed to be coral from the examination of the lead, was passed near Cape Balnc, & so alarmed the Commander, that we saw nothing more of the coast till near Cape Sambu, and then again Cape Nord-ïst, and the entrance of the Straits. A strong current or tide was with us up the coast, and coming into the entrance of the Straits, it turned against us so strong that it was only to be stemmed by steamers. The well-defined shores of the entrance render it easy to make the ship's position as soon as the fog or mist enables [~~the~~] [~~ship~~] [~~to~~] captain to see his headlands.

The boats from the three ships were soon alongside to show the steamers to their berths, and as the harbor opened to view every one was surprised at its security and spaciousness, & the easy access to it. The town lies on the western side of the habor, 25 ri east of Matsmai, and is reported to contain a thousand houses, some of which appear like ware-houses for size, as seen from the ship, and all show better from their position on the slope of the hill than Simoda. The hill rises behind the mass of the dwellings, protecting it on the east, but the land slopes down to a plain on the north of the town and bay, stretching away miles to the base of some high ridges, whose tops are now covered with snow. To the east stretched a low long point, defining the western side of the harbor, back by high land. The snow on these and the hills beyond the straits gave a wintry aspect to the scenery; quite invigorating to us who had just left the warm villages of Simoda.

At noon a party of officials came aboard, with whom Mr. Bent & I went to the Macedonian, where we gave them the letter for the authorities here, which had been written by the commissioners at Yokohama. The bunyo [奉行], called Kudoö Mogoro, had been much terrified by the arrival of the three ships, and in the absence of Namura & Kenzhiro, who had not yet reached the place with instructions, he was utterly at a loss to act, & had refused to see Capt. Abbott, tho' wood & water had been supplied to him. Our explanations & a perusal of the Treaty illuminated their minds more to the purpose, and they seemed gratified at the prospect of intercourse, a meeting being appointed on the morrow on shore to see the bunyo. No tidings of the treaty had reached them, and a journey of 30 days was necessary to come here from Yedo, prolonged or shortened at times according to the season; of course a trip of only four days surprised them a little.

May 18th Hakodade Bay

Four or five of us went ashore this morning, and were received in some state at a sort of public reception room on the beach, the entrance to which was by steps up a stone sea-wall into a yard secluded from sight in the boat by a guard-house. The path across this yard was laid with mats, and a guard of a dozen stood in order to do honor to our entrance into the hall, dressed in blue leggings, swords, and ceremonial jackets. The officials who received us were the four whom we saw yesterday, and politely asked us to be seated on square forms covered with red felt, handing tea, pipes, &c. The room was matted, two sides were partitioned off by screens, & one side was apparently made with closets in the wainscot, as recesses in it were two feet deep; the ceiling was 8 or 9 ft. from the floor, & formed the floor of a loft. In the yard were a few dwarf pines, and a pretty bronze water-jar, a finer piece of such work than I had before seen. After our names and titles were all taken, the three officials came in, and our conference began. The various advantages of trade, houses on shore, liberty to ramble about, and whatever had been allowed us at Simoda, were all recapitulated, and the same demanded of the officers here in compliance with the propositions of the Treaty. The non-arrival of the envoys from Yedo had prevented them from ascertaining the views of the Court, and they wished for time to consider upon our demands and representations, to which we assented till 9 o'clock tomorrow morning, leaving all the papers with them, except one in which they referred to the stringency of their prohibitions. The interview was rather tedious by reason of its having mostly to be written in Chinese, for I did not like to trust to talking, and after settling the hour tomorrow for an [~~hour~~] interview with the

Commodore and the highest functionary here, we proposed a walk, to which they willingly agreed.

Going through an alley by the side of the house, we reached the street, where stood four horses saddled, on which the officials had probably ridden to the house. The street was 20 or more ft. wide, and partly macadamized; the dust had just been laid, and runners were sent before to lay the people too, for on both sides of the street they were kneeling in rows as we passed. The shops & houses were all shut, not so entirely on our account it would appear to keep them warm; but the constant succession of papered windows made the streets look dull. The houses all had a porch toward the street, behind which rose the gable end of the roof, thirty feet from the ground; the roofs were thickly strewed with cobble stones, and each ridgepole bore a bucket of water with a broom in it, which and other buckets in the way, were preventives of fires. No women or children were seen among the crowd, which was not very large, or noisy.

In our walk we went to a large temple, called the "Protecting the Country's Hill", which exhibited a finer specimen of Japanese architecture than we had before seen. The tiled roof rose rapidly fully 60 ft. from the ground, and was supported by an intricate system of girders & posts, resting on varnished pillars; the carving & gilding was superior to anything heretofore seen, & the neatness of the hall added to its elegance, or more properly constituted it. The general arrangement resembled those formerly seen, but on the 6 stone guardians placed in the little shed at the entrance were as many Chinese shaped skull-caps, put on as if to keep them warm, & looking so oddly as to set us a laughing. Another temple, also Budhistic, was visited; it was much out of repair, & like the layer one, had no tablets in it. In some of the temples, the images are furnished with a minbus of copper, & one image of a female had a bamboo, as if a copy of the Virgin. Our stroll took us thro' several streets, & we returned to the landing to go aboard, on the whole gratified with the reception.

In the evening, a few officers took a similar walk, in the course of which they came upon a masked battery of 3 guns, evidently just armed, & probably commenced since the arrival of the Macedonian. They were kindly received by officers & people both, shown into some houses, & no hindrance placed in their way to going anywhere. The town presents a better appearance than Simoda, and the robust people we see proves a healthy climate & plenty of provisions.

May 19th Hakodade Bay

At the interview on shore this morning, the bunyo handed in a long document, in which replies were made to the points stated yesterday, and most of them granted; the paper was drawn up very well, and the dilemma in which he was placed by the non-arrival of orders from Yedo stated, especially in reference to the demand we had made for three houses, which by a singular usage of the Chinese word used they had understood as meaning official residences & court. This impression was removed, and evidently to his satisfaction. The other points were conceded, and after ascertaining the rank of the officer who is to visit the Commodore at noon, Matsmai Kageyu, a relative of the prince, Matsmai Izu no kami, we went back to the ship. At the time, he and the three officers whom we saw, reached the Mississippi, almost seasick with the motion of their shallop, and not over easy at venturing into such a place as they now for the first time saw. I guess their first idea was, at seeing the marines drawn up on deck, that they had been entrapped, but erelong they were put at ease. A copy of the Chinese translation of the Treaty was given them, and the Japanese original handed them for perusal, after which some other points were settled. After the Commodore left, the Japanese remained till evening, and were amused in many ways, greatly to their instruction and quietude, so that when they left, they were put wholly at their ease respecting our designs. The engine, the guns, cables, rooms, and equipment of the ship, were explained as well as they could be, and everything done to make them aware of their neighbor across the Pacific, with whom they were now to come in contact. We were all much pleased with the gentlemanly bearing and intelligence of the two chief men, who were in some respects superior to most of our official friends at the South.

I was told that the Ainos have all been driven or moved to the north of Yeso, none of them living here; their number was stated at 30,000. The Japanese occupy the southern end of Karafto or Sagalien I., and one of the clerks present have been there some years since, glad to return from such a cold, uncivilized region. No coal is found in Yeso, & he took two pieces ashore as a muster. The principality of Mutsu, & Dewa too, on the opposite shore of Nippon, furnish gold & silver; the former is [a] large & rich state.

Sat[urday] May 20th Hakodade Bay

At the interview this morning, the inability of the prince to come here from Matsmai, and the difficulty of seeing him even if the Commodore went there, were expressed in the most decisive terms; while aslo the position of Matsmai Kageyu as his deputy, invested with as full powers as he could have to manage all affairs connected with us, was explained. If the Commodore pleases to think that all this is false, and that he can get the prince to come by ignoring the powers of his deputy, it seems by far the best way for him to go to Matsmai as soon as he likes. The officers here are willing temporarily to allow us to trade, the stipulations of the Treaty showing them that has been agreed to; and today it has been begun in a manner which must rather surprise them, and will doubtless equally please the shopkeepers, as any other course of conduct. The valuation of our dollar, half & quarter was placed at 4800, 2400 & 1200 cash; while a comparison of our gold coins with theirs, made a gold dollar worth only 1045 cash, showing that gold is to silver here is only about 4.7 to 1, a most extraordinary thing, if their coins are of equal purity with ours. Copper must be very cheap, but this does not surprise us like the other. After the interview we went to three places, which were selected for the same purposes as those at Simoda, and also into two or three shops, to explain the manner of trading to some officers we saw in them. The authorities seem to be pacificed and now, their fears alloyed, will, I think, be ready to manage things better than if Namura & his "cross-looker" had come. All this gives me considerable practice in Japanese, and I am in hopes to make the people somewhat acquainted with our character and intentions, and aware that really they have nothing to fear. Some few women were seen today, & more children, but the people have not thought it altogether safe yet to bring their families back to town. It is unpleasant to see how they bow down when the authorities pass by, tho' it should be remembered that custom has made this, what appears abject to us, the natural exhibition of obeisance. In their own intercourse, the officials are far more familiar than at the south, and treat us too very friendly. With the chief man, Yendo Matazayemon, we have become on almost intimate terms, and with Ishizaka Kanzō & Kudoö Mugoro well acquainted; the last is called bunyo, & neither he [n]or the other talk much. Some of the writers are affable, and among all there is a degree of respect and courteousness towards each other & us, which contrasts well & favorably with the people at the South. It is more agreeable too, to see a well-dressed crowd than such almost nude men & loosely attired women Simoda presents. The people here are on the whole larger, I think, than there, and indeed should be, as the climate colder. In a walk thro' the streets, we saw many fireproof granaries or warehouses, and the precautions against fire [&] the fears felt. Provisions are not plenty now, salmon, skate & plaice forming most of the fish bought; crabs & clams are to be had, but not many vegetables. The irish potato is grown here, not in season now, tho' we have got a few; we know not how it was introduced, but probably by means of some of the ships stopping for supplies.

Sun[day], May 21st

Although it was the Sabbath, the usual regulations of the ship we dispensed with, in order that officers of all grades might rush to the shops to seize what they afforded, as if the purchases of a few curiosities was a work of more importance than the observance of the Sabbath, or even the less matters pertaining to the ship's duty. However, Capt. Lee set a better example, & his men were kept aboard, where they heard a sermon from Mr. Jones; and his officers too, were nearly all there. It is strange to see how example in the navy seems to obliterate every idea of a paramount obligation of duty to God; and that it is more important to obey an order of a superior than the plainest dictates of the Bible. In all that appertains to disregard of the Sabbath, the Commodore shows a disastrous exmaple.

Mond[ay], May 22nd Hakodade Bay

The Commodore & two captains went ashore this morning to return the visit of Matsmai Kageyu, whom we found ready receiving us, and mild as usual. The credentials were given from his prince, empowering him to come & receive the Americans and treat them politely; but after he had shown them & made a translation into Chinese, he committed himself by declaring that he had full powers to settle everything; since the question of defining the limits was one he could not settle. We had a tedious conversation respecting it; ten ri was given them as a limit, but this distance would reach to the opposite shore, & therefore 7 ri was proposed as at Simonda, but even this was beyond his powers. He evidently is a man of little energy, afraid of taking any responsibility, and yet gentle in all his refusals, as if desirous to oblige us by assenting. In an hour & over, the Commodore became tired with the slow progress, & gave him till evening for an answer; moving at the same time, to go on a walk over the town. We went to two or three temples, & through the streets, which were quite bare of people, & most of the shops shut. Two or three negroes were standing near a shop, & struck Yendo with surprise, asking several times if their faces were not painted, for he had no idea the kurombo were anything like them. In the evening, we got the same reply that the limits could not now be defined, & also a long paper of complaints against the conduct of Americans on shore yesterday — a heathen prince complaining of the bad conduct of Christians in his town on the Sabbath, gambling in the temples, climbing over walls to get into houses & yards, carrying off things out of the shops, and acting like madmen! Such is a decent moral man, when the restraints of society are taken from off his natural heart.

Tuesd[ay], May 23rd

In consequence of this complaint, all officers were kept on board today, & the matter endeavoured to be rectified, by requiring of the officers & among the petty officers, that any debts due on shore be mentioned, & all swords purchased given up, as it was said this was in particular had been complained of. A lot of presents for the prince of Matsmai, his deputy & the three local officers were taken ashore, & an answer given verbally to the complaint, this morning. It is probable that these officers were alarmed at the rush on shore, & knowing their own dangerous responsibility if anything disastrous should happen, they made the most of the ill conduct which disgraced a few to keep all away. Investigation was demanded, & offers made to return what have not been paid for, or pay all demands. It will doubtless be remembered by the officials & people too, & time only can efface the bad impression now made. In the afternoon, the Commodore came ashore, & took a quantity of articles which had been brought there for his inspection; none of which were very fine, tho' presenting a considerable variety. The odd patterns of silk & cottons are curious [as] anything offered.

May 24th Hakodadi[*sic*] Bay

An effort was made to bring together a number of things for the officers, and by three o'clock Mr. Bent & I managed to induce the Collector to get a broker to bring in a tolerable variety of articles, not nearly so many as we wished, but still measurably gratifying to the purchasers, & in the same degree satisfactory to me, as I was afraid I should not at all satisfy or please. In the morning, Yendo & Ishizuka Kanzo had their portraits taken, and they were hugely pleased to see themselves on the plate, with their retainers behind them holding spears, caps, & bearing their distinctive coat of arms. No one here had ever heard of the art, and curiosity, wonder & delight were about equally exhibited in their manner & questions. The day was good, & the result pleasing to every body.

An answer was returned this morning to the long representation made upon the ill-conduct of some of our ships, in which Perry declared that 7 ri, or 16 miles, must be also considered as the limit, within which Americans might ramble. I think no more trouble will now arise, as the mode of conducting the bazaar seems to give general satisfaction to all parties.

May 25th Hakodade[*sic*] Bay

The shopkeepers in the street, finding that their customers are all going to the fair, have begun to try to better themselves, & to invite customers into their shops, in order that the govt. broker may not get all the profits; this competition of course will improve the market, & call out the goods from their hiding-places, and if it does not again produce trouble, will be an improvement. The goods were much more numerous today, tho' some sorts of lackered-ware were not to be had, and many more people were satisfied; the variety of fabrics was greater, and some pains had been taken to collect a good stock. The seller had a paper before him with the various coins offered all drawn as accurately as he could make them, and placed each one on the drawing to see if the size corresponded, & then compared the effigies.

We paid a visit to Yendo, to arrange about burying a man in the place, who had died on board the Vandalia last evening; he acceded willingly, & soon after went to one temple near by, but no suitable vacant space could be found in its compound. This temple was the Koriō zhi 高龍寺 or High dragon temple, so called probably from the carvings over the doorway, of the two scrambling dragons. Not succeeding here, we went out of town through the seaside-gate, & about a half a mile out, came to an old graveyard, in which a small plat was set apart for the use of Americans. The place is in full view of the harbor, and will contain 25 persons, allowing each room for a tombstone.

There are four large Budhist temples in the town, each of which exhibit the religious zeal of the people in their carvings, gilding, & numerous fine sculptures. The Zhiogen zhi 净玄寺 is by far the most elaborate; the Zhetsu-gio zhi 實行寺 , where Brown's daguerreotypes is well kept, but ancient & inferior; the graveyard near is an interesting place, full of grotesque & handsome monuments, most of them well-carved; the long poles, covered with prayers, standing near them, or lying down, give a singular aspect to the yard. The fourth, the Shio-mio zhi 彌名寺 is old & possesses little interest. In addition to these, there are three large Sintu temples, the Shimmei 神名 , the Hachi-man 八幡 , or the Penten 辦天 , but judging from the second-named & largest, less attention is paid to them than to the Budhist. If there are 7 temples, there are also 7 schools, and girls are taught in them, but I can get no clear idea of what is studied. It must be vacation in all of them since we came, judging by the small number of children seen in the streets. Most of the dwelling & other houses here are built of boards standing up, & made secure by long girders running along outside.

[~~Sat~~], May 26th

The remains of the sailor were buried this morning, and I am able to find only a small stone, on which to inscribe the epitaph, for all the square, handsome ones seen in the graveyards were, I was told, brought from Sado I. & other places in Nippon. The body was brought ashore at the landing, & carried by sailors thro' the streets to the spot designated, numbers of the people lining the roads, all in the greatest quiet looking at the unusual procession, which Yendo himself accompanied to the grave. In all these interments, the Japanese officers have behaved with great decorum, but his kindness of manner has exceeded the others, & no law was quoted by him about looking at the corpse, as was the case with the impudent Isaboro at Simoda.

I spent most of the day endeavoring to get up a bazaar for the Commodore, but did not succeed very well, as in fact the assortment is pretty well exhausted in town. There were some new things, many of which exhibited new features of Japanese art; and many were there desirous to get the articles, as soon as the Commodore had made a selection. Owing to a misunderstanding, he did not reach shore till almost sunset, and found several officers there, (happily most having gone just before,) to whom he expressed some dissatisfaction. He took some articles & went off, whereupon such a grabbing for this & that ensued as was quite surprising to me, and not creditable to naval officers. I was called here & there by natives & foreigners at once, unable to answer half their demands, much less get anything, even if I had wished it. I was ashamed at such an exhibition of American character, in the eyes of the Japanese officers looking at the eagerness and bustle before them.

Sat[urday], May 27th Hakodade Bay

The broker who attends at the bazaar was told this morning that he might sell such articles as he had whenever he brought [the]m there, and manage their sale as he pleased; everybody has had a chance already, & I am desirous of getting clear of the affair. It seems, from the conduct of the shopkeepers, that the broker has taken some means to intimidate them or to prevent them selling much, for it is difficult to get many fine things today, and their prices generally are very much higher, which is no wonder, considering the great eagerness manifested to purchase. Not having any particular business, Morrow & I took a stroll, going out beyond the graveyard, and so on to the end of the peninsula on which the town lies. We enjoyed the walk very much, found many plants, and saw a few people only. Some of the plants were old acquaintances, especially a Trillium, a Viburnum, an Anemone, a Mentha, and others, growing naturally in the woods among the bamboo, a small species of which is common here. The extent & variety of seaweed here is great, and vast quantities are used for good by the people. On our return, we went thro' the grove of pines & cedars behind the town; a delightful place it must be in summer for the townsfolk to ramble in. The hill-top affords a fine view of the surrounding country, and the patches of snow on the western peaks showed us the latitude we were in. No terraces, such as are seen at Simoda, are seen here, and the plain north of the town, is neglected, naked, and almost uncultivated, & the pursuits of agriculture occupy only a small portion of the inhabitants. The country is not thickly settled in the immediate vicinity, and most of the supplies are brought from the south, Simonoseki, Sado Is., Yechigo, & Ohosaka, being the chief ports, from which not only rice, wheat, cloth, porcelain, lacquered ware & cutlery are brought, but also common things, as gravestones & tiles. What these imports are paid for with, I have not been able to learn.

The Commodore made some inquiries about shipwrecks on the coasts of Japan today; and at the same time, invitations were given the officials to visit the ships, if the weather was fine on Monday. The more I see & am able to talk with these [~~officials~~] the more favorably do they contrast with the same set [of] men at the south.

Sund[ay], May 28th

Early this morning, I was sent ashore to inform Yendo of the death of another seaman on board of the Vandalia, G.W. [R]ick by name; he expressed a good natured sympathy with the death of a young man so far away from home, and pointing out a new wharf to land at just above the Commodore's house, where he wished the body to be brought on shore in the afternoon, asked if he should accompany the body, to which we, Mr. Bent being with me, said that it was not required by any of our usages. In the day, Mr. Jones delivered a practical discourse on the first clause of the Lord's Prayer, which made one feel too that to the Japan the same Father extends his care, & I hope will erelong send the evangel of salvation.

May 29th Hakodade Bay

The rest of yesterday's Sabbath was pleasant in the extreme, and I was willing to begin again this morning. I shall not be so much harassed this week as last, for now trade is carried on at the custom-house without my assistance. The officers & their friends were invited aboard the flag-ship to visit her, and then to go over the Macedonian, and spend the evening in seeing the performances of the Ethiopian Minstrels. All came but the Prince's deputy, Matsmai Kageyu, who had a bad cold, and left for shore after 9 p.m., much pleased & diverted with the show and the Commodore's entertainment, which was got up remarkably well for the means & time at hand. A lot of presents were also brought at the same time in return for those sent by Perry to them, paper, umbrellas, crapes, dried salmon, fresh fish, &c., altogether worth about a rifle & a pistol; to each of them had been sent a rifle, cavalry sword, pistol, box of tea, 12 btls of s whiskey, 12 ps. cottons, perfumery & cherry cordial.

May 30th Hakodade Bay

This has been a quiet day, for after seeing Yendo to stir him up about the accounts of the ships, and getting the answers respecting wrecked vessels, I took a pleasant walk with Dr. Gilliam after flowers, & went back to the ship to dinner, one of the few times I have had a good opportunity. In the afternoon, I had to wait so long for the accounts of the Vandalia & Macdedonian, that I had time only to close up a letter for Canton by the former, & send Dr. Bridgman[38] his book & some India Ink. The weather has been so cold today as to make a fire comfortable; the climate must be much colder here than on the same latitude in U.S., where no snow can now be found on hills no higher than those hereabouts, the highest of which may perhaps be 3500 ft. and not bare on the summit, so far as his glass can decide.

Wed[nesday], May 31st

My commissions multiply apace, as I am requested by one after another to procure things for them on shore, most of which are not to be had; however, I was able to get some things for Maury & Maxwell today, which showed at least my good intentions, if I could always succeed. An effort was made to get a block of stone here to take to Washington for the monument, which Perry wishes to exchange for a map of that city. The "Vandalia" & "Macedonian" were out of sight before 10 o'clock leaving only a small show of two vessels in the harbor. This evening we learned that the commissioners had come from Yedo, and would be ready to see the Commodore tomorrow. They have come so lately here that it is not yet time for them to learn what has been done, [~~here~~] & it is rather too late for them to undo it. Mr. Bent & I went to the graveyard this evening, & found that a substantial fence has been put up in front of it. I got a shingle epitaph too, with a Thibetan inscription on it, & Mr. Bent procured a dog, for which he had some difficulty to pay the man at the custom-house; when he had been forced to take the money, he went away & erelong returned with a pair of white ones, which he made Mr. Bent take; & before the boat left for the ship, five or six were brought down for [one or two words lost] taking off. The breed here is like the Chinese. Horses sell [one or two words lost] $25, for common hacks, & $300 or more for barbs.

June 1st, 1854 Hakodadi Bay

Six years today since I left New York, and now I am thus far from that city and on the journey of life.

Early this morning, Tuzhiwara came aboard with a note from [Amma] Zhiunoshin and Kenzhiro, announcing their arrival at Hakodadi, en route to Karafto, whither their superior had already gone, and expressing a desire to meet the Commodore, and that the business in hand would not detain them more than three days. It was agreed that we should go ashore at ten to fix an hour for them & their suite to come on board. When we reached the house (Yamado) the hour of 1 p.m. was agreed upon, & I was left ashore, while Mr. Bent went back to tell the Commodore. Meanwhile, I started off with Yebiko Zhiro to find a suitable stone for the Washington Monument, and fairly tired him out in the search. We went to the fishing-hamlet of Shirasawabi east of the town, but none suitable could be seen around it. However, I saw specimens enough to induce me to get him to go up towards the hill where the stone is quarried, but it was another thing to get him to take so long a walk. Near this village, most of the gravestones were covered with mats to preserve them from the effects of the frost during the winter; and the graves had just been swept and trimmed up, it being the 5th of the 5th month yesterday. The village was noisome from the drying & decaying fish in it, and I was in a hurry to get away. Pursuing our walk for a mile in the direction of a path, which led up the hill, I came to a couple of long stones, of red trachyte just dug out, and had them marked, much to the gratification of my companion who was fairly used up, or else vexed. While we were so warm on the walk in the noontide sun, the sight of the snow lying on opposite hills was rather tantalizing.

On reaching the house at 1 o'clock, the officials requested us to go to the other landing, thence to take the dignitaries on board. Now the new ones reached the place, however, till three o'clock, and as they would not go off without Kenzhiro, we went away without them. It was a curious sight, as these officials were announced to be on their way to the house, to see the attendants and common people arrange themselves along the path, squatting down with caps, staffs, & other insignia in their hands, bowing their heads to the ground under the effluence of power as it swept by them in the persons of these men, & stood near the landing, but they paid us no notice, as they went into the house, Yendo escorting them.

We had waited now so long, that on the way back we met an order to return; and found the Commodore in high dudgeon, which we hardly had anything to meet by way of explanation. He ordered the marines in both steamers to get in readiness, & 100 blue jackets to land in the morning with two field pieces, in order to show the Japanese that he was not to be trifled with. About 4½ o'clock, Amma, Kenzhiro & others, with the local officers Yendo at their head, came aboard; the Dutch interpreter, Takeda Ayasaboro, had written out a few sentences, stating that he was able only to write Dutch, & could not speak it. He was the tallest and one of the best looking Japanese I have seen. The Commodore thought best to accept their apology, that it was owing to delay in the preparation of a present which had detained them (a good commentary on Prov. 18:16) and they were taken down into the cabin. The conference came to very little in addition to what had been before discussed with Yendo, and the final settlement of the limits to which Americans can ramble in the region of the town was referred to the commissioners. The disappearance of the women & children was ascribed to fear of us on our arrival, and this was now wearing away. The conference was slow [~~but~~] [~~kindly~~], and the visitors from Yedo were gratified with the sight of the ships, Kenzhiro remaining in the cabin with the others went over the decks. He said the journey to Matsmai had been tedious and also slow, often going only 12 miles a day, snow, cold, rough and weariness, being among the discommodities of the way. I suspect these Japanese officials endeavor to live such easy lives, that when they are obliged to go thro' hardship, they suffer much. From his white underdress, we learned that Takeda Ayasaboro belonged to princely blood; he seemed to be often referred to by Kenzhiro, who hardly ever asked Amma about anything. His position & learning probably got him the place of interpreter. Before leaving, the Commodore told them he would return their call in state as at Yokohama, a proposition which pleased them all, especially Tuzhiwara, who was glad to hear that the "so duda do" were coming ashore in their fine dresses.

June 2nd 1854 Hakodade Bay

A fog soon bedimmed the prospects of a fine day, and before ten o'clock Perry had decided not to go ashore, which seemed to be the most judicious course, as the fog seemed likely to condense into a rain. Presents of a sword, rifle, revolver, perfumery, tea & whiskey, were prepared for Amma & Kenzhiro, and an apologetic excuse to explain his non-appearance. Mr. Bent & I found the house in readiness to receive the party, and an unusual attendance of servants showed that some preparations were making for the occasion; and tho' the Japanese apparently took it in good humor, their disappointment was evidently great. Along the street, too, were many signs of the expectation of a gala and fine show among the people. It need not be said, in what a pet the Commodore and most of the officers would have been, if the Japanese had excused themselves from an interview for what appeared to be such inadequate reasons; and how many denunciations we should have heard. The presents were handed to Kenzhiro, but the answer to the Commodore's note was not ready. The purveyor's bill was paid as follows: for the "Powhatan"

100 bskts charcoal 5.25
100 pine boards, half in. 1.38
50 pine boards inch. 6ft. long 0.80
100 pine boards inch. 20ft. 4.67
285 lbs. Sugar 7^4 19.85
1350 sticks of wood 10.69
500 brooms 3.65
Total 46.29

1000 sticks for "Southampton" $7.92
976 sticks for Mississippi 7.71
15.63
6733 sticks for "Macedonian" 36.52
1891 sticks for Vandalia 18.20
54.72
Provisions furnished Powhatan 34.12
$150.76

[June 2nd Hakodadi Bay]

The prices of these things were repeatedly declared to be equitable & sufficient, but they were so low, that the Commodore made the purveyor, Inagawa, a present of a box of tea; & the boatmen a larger one of 900 lbs., biscuit, 3 bbs., beef & port, & 60 lbs. tea, for their labor in bringing wood & water. With this all parties were satisfied. The two blocks of red trachyte came off in the afternoon, when a further small addition of provisions was made. Two of us went ashore to obtain the answer, for which we had to tarry till nearly sunset, when we took leave of the friends, whom Mr. Bent & I had become quite attached to during the fortnight as had been in port. They also evinced very different feelings from those apparent at the first interview, and we parted with expressions of mutual goodwill. Three of the newcomers, Takeda Ayasaboro, Yushimi Kennozhio, & Tsuji Kayemon, came off to the ship with us to see it more closely, and remained until it was too dark to see anything; they evinced considerable knowledge as well as curiosity, especially the first, whose acquaintance with Dutch had opened to him sources of information not accessible to the others.

June 3rd [Hakodadi Bay]

Early this morning we were under weigh, but the fog came in so thick that both vessels came to anchor; and soon after, a boat came alongside with Yebiko & Daishime, to see why we had returned, supposing some accident had happened. They remained on board till we were ready to start, much interested in the appearance of the machinery in readiness to be put in motion. Thus ended our visit to

Hakodadi, forming one of the pleasantest episodes in my life in Asia. I expected a dull visit at a miserable fishing village, while I found my time and abilities employed to their highest degree, the whole business of interpreting thrown on me, & the duty of removing from the minds of the officers their apprehensions, and disinclination to act in the absence of orders from Yedo. Acquaintance produced mutual trust, and as they found themselves fully supported by the treaty, it was soon seen that no little trouble would be avoided by meeting all our reasonable propositions. It was favorable to them that the lack of particular instructions from court left them more at liberty to follow what the Treaty implied; and it was more favorable to us that we had two such persons as Matsmai & Yendo to deal with, instead of two petty minded & hestitating men, like Kondo Riozhi & Tatsnoske at Simoda. I have been repaid, during the last fortnight, for the years of study of this language; & hope that the impression left at Hakodadi may by & by open the way for a residence there to some one who will tell its people the knowledge of the love of Jesus for their souls.

[names of high officials in Hakodate, written in Chinese characters, with their respective English translations as follows:]

Of the family of Matsmai Idzu no kami, was 松前伊豆守家来

Matsmai Kageyo, a high minister. 大夫 松前勘解由

Yendo Matazayemon 用人 遠藤又左衛門

Ishizuka Kanzo & 町奉行 石塚官藏

Kudo Mogoro, (the bunyo) 箱館奉行 工藤茂五郎

Local authorities

Sheki Nakaba 関央

Fuzhiwara Shiyume　藤原主馬
Daishima Gohei　代嶋剛平
Yebiko Zhirō　蛯子次郎
Clerks & secret spies

June 7th　Simoda Bay

The passage hither occupied just 100 hours, fully 15 more than it wd. have done if a thick rain yesterday afternoon had not made it, in the Commodore's opinion, unwise to go to the west of Oö-sima.

During the night a current carried the ship southeast & south so that we did not anchor till nearly 1 o'clock; the weather turned into bright sunshine today, showing the green hills, with their naked summits and patchwork of reaped and ripe fields of grain, adown their sides, in pretty contrast. The stimulus of rain & sunshine has made surprising improvement in the face of nature here, since we left it 25 days ago. The commissioners are all here, one load of coal has come and part of the supply for the bazaar. We went to see the prefect in regard to an early interview, which is to take place tomorrow.

June 8th　[Simoda Bay]

According to previous agreement, the Commodore landed today at noon, under a[n] salute of 17 guns, with as large as an escort the ships could muster, composing a force of marines and sailors with four fields pieces, numbering in all, including officers & musicians, upwards of 300 men. The day was unimpeachable, and the way from the landing to the temple was lined with the people, whose talking, as we moved on, was not unlike many beehives in commotion;

so that above & below all combined to make it interesting to all parties. It was very different indeed from the visit paid by the Russian ambassodor Resanoff to the envoy at Nagasaki, when the people were kept away, & all the streets lined with curtains to hide even the houses from the view of the Russians. The music sounded gaily as the line passed into the yard of the temple, and the whole formed an excellent subject for a painting, when seen from a favorable standpoint at this moment. On entering the yard, the Commodore was received by Kurokawa, and conducted into the main room of the building, which had been so transformed & divided off by curtains & folding screens, that it was not easy to recognize its former appearance—a use which shows that the Japanese apply their religious edifices to the same general uses that they do in China. In this main room, stood the five commissioners, with Hayashi at their head in scarlet-trowsers, and two additional ones, who have been appointed to the body. We were conducted into a side room, and the two parties seated opposite, just as they were ten weeks ago at Yokohama, except that Mr. Bent had taken the place of Capt. Adams. The discussion was tedious, continued for three hours, and only a part of the subjects introduced decided on. Lin wished to put up guard-sations at the limits prescribed to the rambles of Americans in the region of Simoda, but Perry wished to have it previously ascertained that they were not within the 7 ri agreed on by the Treaty; and deputation is to visit these spots, & then report. The decision of the limits at Hakodadi was also more difficult than we had supposed it would be, for the Japanese were not ready even to make it the same there as at this place, nor to propose any distance themselves.

They wished, however, to get the Commodore to take

away the big box he had placed on the southern side of the entrance to the harbor, and also to remove the buoys over the rocks. The only explanation we could give for such a proposition on their part was, that they had construed these proceedings with reference to some idea of our thereby taking possession of the harbor, or at least, driving a nail in that direction. He properly refused to remove the buoys, and suggested the appointment of pilots before the box was taken away, who could show ships the dangers it cautioned them against; and they agreed thereto. After this, the drawing of the Washington Monument was shown, and the proposal made them to furnish a stone to put into it, adding that one had been procured at Hakodadi. These discussions, and a collation of cakes & fish, filled up three hours, when the session was adjourned. Before leaving the temple, the marines were marched & drilled, & the manner of using the field pieces shown, greatly to the satisfaction of the Japanese. The Commodore & his suite returned on board, but the men were marched down to Kakizaki, followed by a large crowd; it was a gala day to all parties, except Lo [羅森], who got quarantined for not coming off when the ship's boats came back.

June 9th Simoda Bay

The slow progress yesterday induced the Commodore to send us ashore this morning to have a talk with Moriyama beforehand, in order to hasten matters to a conclusion; but it did not apparently have any effect, for the Commissioners had their own matters to bring forward, some presents to spread out for acceptance in exchange for those received, and arrangements to agree on respecting valuation of coins and party to go & settle the limits of 7 ri. How droll those seven bald shaven men look stretched along in a row, as they sat opposite me today! Lin in his scarlet browses, & the silly, vacant-faced Matsusaki, one at the one end looking grim and dignified, the other at end, sleepy and silent. These interviews are instructive, too, taking into account the circumstances under which we all have been brought together, and the Japanese officers seen qualified for their places, in the main. Some presents were given to Tsudzuki, Prince of Suruga, & Takenoüchi Sheitaro, the two new commissioners, rifles, swords, perfumery, &c. Some of the articles sent in exchange for the howitzer were finer specimens of manufacture, mostly lacquered-ware, & fully equal to it, taking them all, in value. The conversation to-day was more general and pleasanter than we had before, touching on many topics. We learned that the four first commissioners are all merely titular princes, & have no authority over the principalities they take title from. Moreover, that there are over 500 athletoe in Yedo alone, & hundreds in Ohosaka, all of wrestlers get a living by exhibiting their prowess; yet I think that the strongest one among the ninety we saw at Yokohama would not prove a match for some of the boxers of our country or England. They eat little or no meat, and develop more fat than brawn.

June 10th Simoda Bay

Mr. Maury, Bent & I went early this morning to see Kurokawa respecting our trip to define the boundary to which Americans are permitted to go by the Treaty. We are received at the temple, & the matter seemed fully understood on all sides. After a while, Ido & Izawa sent in word they wished to see us, and soon appeared themselves, expressing their pleasure in polite terms, and giving each of us a piece of silk for our wives, and four stone bottles of saki and a box of sugar-plums to beguile the wearisomeness of the way in the journey of today. So much for their hosptable intentions; and we went aboard to get ready for the terrible jaunt they had described. At noon, we were ready at the landing-house, with attendants, instruments & baggage, but saw nobody ready there to take the latter, or signs of much preparation on the part of the few Japanese officials thereabouts. We got them to start in half an hour, however, and proceded beyond the temple through the stone-cut gorge to a station-house at the foot of the hill, where we were desired to stop, for this was one of the guard-stations defining the limits of the jurisdiction of the governor of Simoda. It now appeared that there was a mutual misunderstanding, for the officers said we would now go to the next guard station, while we said we wished, & were ordered, to go to the end of the seven miles. Isaboro & Tatsunoske soon arrived, and told us in no less plain terms that the commissioners had no idea of our going beyond the guard stations, & no preparations had been made to lodge us. Mr. Maury sent a note to the Commodore, desiring instructions, & we went on, followed by our cortége. The incident was a good illustration of the easiness with which confusion of purposes may arise where the medium of communication is so imperfect, and little pains taken to state the intentions of each side. Isaboro accused me of misinterpreting and lying; so [~~that~~] Mr. Bent was addressed in a long speech in Japanese, & to make the matter plainer, Tasnoske tried in vain to put it into English. They both returned with the Commodore's reply by which time we had reached & passed another guard-station and seemed glad it was now cleared up, tho' I did not see wherein their responsibility consisted. We crossed over a number of hills into the [~~town~~] hamlet of Hongo, where the station is to be placed, & returned to Simoda at evening. During the interview today, some matters were settled & others brought up which last showed the fears of the commissioners lest they had or should give us too much liberty. From the general tenor of conversation, we gather that they had been blamed for allowing so much extent of rambling as the treaty states.

[Sunday], June 11th Simoda Bay

It rained during the whole day, so that there was not only no religious services, but no coaling ship either, which it was intended should occupy the Sabbath in both steamers. Consequently, there was some rest for the men, though orders came for them to resume coaling at sunset. The Macedonian returned this evening, the Southsampton having been in two days. The latter had a misty spell of weather at Volcano Bay, but Capt. Boyle was able to make a survey of the harbor, and go ashore a few times. The Ainos & Kuriles were more numerous than the Japanese there, but lived in a most wretched manner, destitute even of the comforts of the Japanese, subsisting almost wholly on the products of the sea & hills, and under the complete sway of the Japanese. The antlers of deer were common on the ground near their houses, and some [~~were~~] deer were seen on the hills. They were very hairy people, as described by La Peyrouse, and with their scanty garments, such additional covering would be comforting; tho' I would not say, as Lamarck would, that the hair on their backs grew two inches long, because their jackets were so thin.

June 12th [Simoda Bay]

The conference this morning was more tedious than ever, and small progress was made. The Commissioners refused to let a party go to Oho-sima, nor would they consent even to $3^1/_2$ ri as the limit of rambling at Hakodadi, less than which Perry declined to consent to. Three pilots were introduced, like spaniels on their four feet, to whom the business of conducting ships into the harbor was to be committed, and no pay was to be taken for this service; in this manner the government will have their spies on board on our ships before anchoring. The project of going to the limit allowed was discouraged, but its introduction brought out the suspicions entertained lest we should remain on shore over night, and the Commissioners seemed to think no Americans were ever likely to need to sleep in Simoda, notwithstanding the Treaty made provision for a consul. Of course it was disallowed, & they were told that they had better set up ten houses or taverns for the accommodation of seamen, than to try to keep them thus on board ship. While we thus the discussion [*sic*], reports came in of misbehaving, and on going to the landing, Perry found some of his bargemen and bandsmen so drunk they knew not what they were doing[;] a couple of bracelets met them on board, but it was a bad corollary on our discussion. Simoda, like Canton, is likely soon to have its Hog-lane; and the worst feature of heathensim & Christian nations, exhibited, making human nature more repulsive, before the excellencies of Xty come to be known.

June 13th Simoda Bay

Mr. Spieden & Mr. Eldridge took me along with them

this morning to assist them in the discussions respecting the currency in which there is likely to be no little difficulty, arising some degree from the mistake we made in offering to value our dollar at 1200 cash, & letting it go at that until we went north, but still more from the evident desire of the Japanese to force us to pay in our gold & silver at their arbitrary valuation. On reaching the temple, we found Kurokawa & the committee ready to meet us, clever people sitting in solemn rows to take note of what we & each of them said. Setting aside what was done yesterday, we began by proposing an equal exchange of gold for gold & silver for silver, and after no small delay, made them produce two ichibu [一分银], whose weight we compared with our dollars; they agreed that three ichibu made one dollar, but refused to consent to an exchange, saying that their valuation of gold & silver was so arbitrary that no reference could justly be made to it in conducting trade. It was twleve o'clock when we had reached this point, & the Commodore came in, rather surprised that in three hrs. we had made no more progress. At his session, which lasted till six o'clock with only a short interruption, the limits at Hakodadi were settled at five ri [里], tho' yesterday he offered them $3^1/_2$ which they would not accept, & they had before offered five, which he declined. The temples at Simoda & Kakizaki were offered as places of resort for the sailors, and the desirableness of establishing shops or inns was urged; and what was characteristic of Japanese & Chinese sway, Lin desired the Commodore to give orders that no sailors should get drunk on shore as they did yesterday, as if this was our responsibility. Perry told them, this was their lookout, & if the Japanese did not sell sailors saki, none of them would get drunk. A complaint was made against one officer for leaving religious books at one of the temples, upon which the Commodore said that if they would point out who had done it, & bring back the books, he would give orders in the matter. He then said, that if the priests at the temple had not willingly taken the books, none would have been left there; and made a complaint in addition against the obscene books which the Japanese had given the sailors, & thrown into the boats, declaring that such things were worse. He said that the Americans have no desire to interfere in the religious views of other nations, perfect freedom was allowed in those matters in U.S., where even the Japanese might have a temple if they chose; but that they would never suffer the Japanese to insult the Chrisitian religion, and any attempt to cast reproach on it would be met with opposition, and bring down on them the anger of the American people, wherefore, it would be well for the Japanese to treat Christianity with respect. — Another point they tried to get Perry to consent to, the accompanying officers with spies under the name of guides, attendants, interpreters, or servants, was rejected, & the entire freedom of Americans to go as they pleased within the limits, staying out over night even, was maintained, as being granted in the Treaty. A letter was brought in, just recd. from Hakodadi viâ Yedo, inclosing some of our written conversations held there and stating that [~~if~~] Perry had declared, that if he could not have ten ri about Hakodadi as the limit, he would make the Japanese pay 10,000 cobans as damages. The matter was placed in its true relations, but I could understand enough to hear them charge Lo & me with misinterpreting on these matters, and making trouble.

June 14th [Simoda Bay]

The finance committers separated today, unable to come to any agreement, for the Japanese refuse to exchange our coins at the value in cash of silver, but regarding our dollar as bullion, they give the nominal valuation at the mines, where [~~its~~] weight is reckoned by taels & mace, and cheat us of just $66^{2}/_{3}$ cents in every dollar. The currency is now perfectly arbitrary, for the toö-hiaku is probably not worth more than ten copper cash, while it goes for 100; and compared with silver it is as cheap again as our cent, being nearly four times as large & only rated at 2.05 cts. Silver compared with gold is actually about 4800/1045 ths, or $4^{3}/_{4}$ times only dearer, but discarding weight for weight, supposing an ichibu as pure as a gold dollar, the prescribed valuation makes $20 worth $10.45, whereas $20 silver would be worth $6.66, or an ounce of gold worth $8.448, & one of silver 33cts, or 25.6 times cheaper. This most extraordinary valuation was acknowledged as forced upon the people by their rulers, but the latter would not take our dollars by it, pocketing the difference. If we disliked these terms, we could stay away & not trade. In giving gold, however, when compared with the prices paid by the people in cash, it must be despreciated as silver, and therefore actually is worth only 17 ct., making our $ when compared with the rates of currency among the people, worth $3.45! Yet the Japanese actually make five times a greater depreciation of our silver than gold, for while the latter is as 22 to 17, [~~while~~] the former is 33:100, so cheap is gold here compared with silver. Of course, we refused to agree to any such depreciation of our coins, and broke up the conference. In the afternoon, the additional regulations were agreed upon with Moriyama, he starting out stoutly for discarding entirely the use of Chinese in all official communications, evidently, I think, so as to keep the whole intercourse in his own hands; it was compromised by allowing no Chinese when there was Dutch interpreter.

June 15th

The draft of the Regulations was agreed upon today. They refer to guard-houses, pilots, public-houses, mode of purchase, articles, limits at Hakodadi, and such things. The corpse from Yokohama was brought down today & interred by the side of Parrish at Kakizaki, the Japanese behaving very kindly in the matters. The weather is getting now very warm 75° or so, the wheat & barley are reaped, and vegetation appears thriving. Irish potatoes are cultivated here, and will furnish good supplies to ships if raised in quantities.

June 16th [Simoda Bay]

A third conference took place today between the parties in sessions upon the Regulations, which completed them. In the evening, concert was given on board of the Mississippi by the minstrels at which fully 300 Japanese & 500 foreigners were assembled, mak[ing] altogether a very respectable audience. The ship was dressed up & the dinner was, considering our means, very good; the seven commissioners & three bunyos all sat down, leaving room for only a few officers, the rest being entertained on deck; everything went off well, and no fault could be found with the performances, which were more spirited than at Hakodadi. The only drawback was a slight rain, which incommoded us all during the signing but nearly ceased before the party separated at about 10 p.m. The Japanese were exceedingly amused at the dancing & tambourine music.

This entertainment, and the similar one given at

Hakodadi, will I think produce the impression which we desire to make that we are willing to make all the efforts we can to please the people who have done almost nothing of that sort of thing for us, not even to invite us to a common entertainment or amusement of any sort, or to go & see anything. The commissioners have shown themselves reserved on every point relating to the promotion of good personal feeling, confining themselves to official acts only; and the Commodore has set them a good example. The Japanese hardly know how to behave towards foreigners; they have been so long shut out from them, that both officials and commoners are afraid of overstepping some regulation, whatever they do. This, in some measure proceeds from fear, but a good deal more from haughty pride & contempt of others; the mutual ignorance of each other's langauge further opposes much intercourse.

[One sheet of the printed document is inserted, concerning the exchange rate, dated U.S. Frigate Powhatan, Simoda, June 15th, 1854, reported by Wm. Speldden and Joseph C. Eldredge, Pursers U.S. Navy, to Commodore M.C. Perry. This could be an example of whatever printing SWW did on board ship, using the expedition press.]

June 17th Simoda Bay

The Commodore sent his usual quartette ashore this morning to see the officials about the accounts and the stone and bazaar, and what not, but we made very little progress in getting anything, and the latter seems likely to prove a failure. The Japanese have not half the business tact which characterizes the Chinese, & more especially do matters of trade move slowly when the officials get hold of them. At three o'clock the Commodore went to see the officials, & exchange the triglott copies of the Regulations, but they were not ready nor were his sealed, and therefore no exchange was made. They expressed themselves greatly gratified with the performances of last evening, and were so doubtless. It was not till nearly six o'clock that we could get off, by which time it was too late to think of taking a walk. The harmony of our conference today was marred by two of our crew going into a shop, pulling the spigot out of a barrel of saki, & drinking a basin-full of it, letting the rest run on the floor meanwhile; as the owner tried to stop them, they drew on him, & wounded him in the hand, themselves too being somewhat mauled in the scuffle. Such is one of the precursors of the trade with Christian America, tho' I hope the Japanese have discrimination enough to perceive & make a difference between the sailor who behaves & those who act like friends. It is amazing to see the lengths the thirst for rum will drive a man, and five or six fellows are constantly at the stanchion for their misdemeanors growing out of love for liquor. The officers love it almost as well, but take their own time when to have a bout. Yet there are some of the latter, as much in love with gambling as these are with grog, who pretend to look down with disdain on the poor sots. Altogether, my ideas of naval life will suffer a great depreciation by this cruise.

June 18th

The Commodore moved aboard the Mississippi again this morning, about 14 mos. since he left her. The chaplain had service, but no sermon; and as one might expect, there was not much quiet on board during the day, while there was a great deal of trading on shore. Truly may it be said that life in a man-of-war is too often like living on the outskirts of hell.

June 19th [Simoda Bay]

Today was so stormy that nothing could be done, and the bazaar was deferred by Com. Perry, as he himself was not desirous of going out in the rain. The articles were laid out indeed, but not marked, and we had them all labeled and their prices given, which at only 1600 cash to the dollar were exhorbitant, making the greater part of the articles twice or thrice as dear as at Hakodadi; moreover the variety was much less than we had been led to expect, deficient in many sorts of things which we had learned were abundant in Yedo, and not satisfactory in any department. The bad policy of their persisting in this unjust depreciation of the silver we paid them was again shown them, but either there is some reason why they had rather risk the loss of all trade, or the establishment here is placed on such a footing that it must have this high commission for managing it, and they will not change. The Commodore expressed his indignation at this mode of doing business, that it was wholly opposed to their professions of friendship, and that he would have nothing to do with the matter if they did not change, and make the prices of silver & goods more comfortable. However, there is no likelihood of any modification.

We made some propositions respecting pilots & prices to be paid for them; also concerning some spars ordered by the Commodore, which we were cooly told were still growing in blissful ignorance of their fate on the mountains. In fact, these officials have become tired of supplying our reiterated wants, which, with the provisions consumed by so many of their own officers, must be not a little troublesome, and perhaps experience too, and not worth doing too much for.

June 20th Simoda Bay

The replies and delatory actions of the Japanese were so unnecessary and impertinent yesterday, that the Commodore quarantined the officials from going ashore at all, and sent a document to Lin and his colleagues, showing that they had violated their promises in respect to furnishing supplies and procuring articles wanted for the squadron and himself, especially in some dresses and the spars spoken by yesterday, adding that they were acting foolishly in their own view by not trying to do more to show their professed regard for the Americans, intimating his own opinion of such a conduct and of the power he held in his hands. The paper was put into Dutch (no Chinese now being used in our intercourse), and given to Moriyama. How he rendered it to the Commissioners we do not know, for he has the throttle valves of our intercourse in his hand, but in the evening he came off, and said that the non-procurement of the dresses was his fault, & [that] of the spars was owing to Tatsnoske's carelessness, as he had failed to attend to them. I suppose, that at Desima, no care for such requests even fell to the lot of either of them, and they gave themselves little concern about them here.

After this message had been delivered, we made excuses for Perry's not coming ashore, which were mixed with as much moonshine as usual on such occasions, and I suppose received by the Japanese in a diplomatic sense. They however gave us, Perry Jr. having gone aboard to report progress, the dinner which had been prepared for the Commodore, by far the most elaborate entertainment yet provided. It was served up on small lacquered tables, and a set of little lacquered bowls and chinaware plates, the large articles being brought in on bowls & charges, and served out to each person by the prefect and his aids. Warm and cold saki was offered, the former in thin cups of porcelain brought in floating on water. Less fruit was introduced than among the Chinese, and no candy or sweetmeats. We made the entertainment pass off as well as we could, but both parties felt rather awkward, feeling that it lacked its chief objects, neither Lin or Perry being there. After dinner, a variety of little articles were brought in as presents, not alone for the Commodore, but his suite & Capts. Lee & McCluney. In the exchange of presents, the Japanese have not shown themselves at all generous, whether it is owing to their entire ignorance of the actual cost of the things given them and therefore inability to judge what would be of corresponding value, or to their petty characters. We stayed ashore till 2 o'clock, and I then went to see how the tombstones were being put up at Kakizaki, and found that the Japanese are very expert in stone-cutting, but the material does not retain the inscriptions for many years. They have customs quite different from the Chinese in their rites of sepulture, one of which is cremation, as was seen hereabouts a few days ago. Among other events of today, was the delivery of about 16 tons of coal, which the engineers decide against, even at the price of \$$27\frac{1}{2}$ per ton, and of ten or 12 cords of firewood, a large part of it sticks from 1 to $1\frac{1}{2}$ inch in diameter. These important supplies are therefore not so readily furnished as was hoped they would be, and are inferior in quality. Perhaps a [~~more~~] constant demand may increase the quality, as well as the quantity, and this will probably decrease the price. An exchange of cottons or other goods will doubtless make an opening for the barter of other Japanese articles.

June 21st Simoda Bay

The quarantine continued till three o'clock today, at which hour the bazaar opened. The Commodore sent Mr. B. & [~~I~~] me ashore early, to make the arrangement for exhibiting the things against the time he landed, but when we reached the temple, Kurokawa & Yenoske showed plainly that they were in high dudgeon, & that the scolding document of yesterday had made them angry. The prices which had been attached to every article yesterday had been taken off, and they proposed that, except a portion which had been set apart for the President, the remainder should be taken off to the ships at such prices as we pleased to pay for them. It was with much intreaty and explanation that I got them to alter their minds, and restore the labels, and put their own prices upon the articles, declaring that the Commodore otherwise would not take a single thing nor allow the officers to buy, much more disallow them to be taken aboard ship. After some hesitation & talk among themselves, they came around to our views, and began to restore the labels and spread out the articles. Those for the President were mats, dresses, shell-work, plants and various birds. By the time we had made these arrangements and began to number the goods and list them, Perry arrived, so that there was no need of saying anything respecting the matter. He chose [~~out~~] nearly a hundred dollars worth, and had them sent off to the ship, by which time the Commissioners were ready to meet at dinner. The two chiefs were seated opposite for the last time, but Lin has not much conversational power, and the others, especially Takenoüchi, took the lead in talking. The construction and use of pistols & cannon & steamers formed the main topic of conversation, though now and then other points came up. The interview was a pleasant one, and I could not but pray God that the officers of this hitherto secluded land, of whom so fair a representation sat before us, might be guided by him to change their views and policy in accordance with the new state of things now coming upon them and their country.

The feast was no better than that given us yesterday, and lasted about an hour and a half. We were only interrupted once, and that was with the usual errand, by the orderly in waiting, telling the Commodore that the bargemen had run away into town, doubtless to get spirits. After leaving them, Perry went aboard, and we made ready for the coming of the officers. The numbers were rolled up, and put into a box, Mr. Perry giving them out; there were nearly enough to go around twice, and as is usual, the coveted things were drawn by those who least expected them, Mr. Caulk, the gunner of the Mississippi getting the large paper-box. However no other way of getting the few fine articles distributed without dissatisfaction was available, and there were enough in all to let each officer get something. It was a busy time for me for about an hour or two to get the various articles drawn for by one & another, ten or twenty of whom drew what they could not find. Before night, there was very little left unsold, a part of the umbrellas, shoes and coarse baskets remaining, only; while ten times as much fine lacquer could have been disposed of if it had been there. The assortment was far less than we had expected, and I think far less than any Japanese merchant would have produced, if the affair had been instructed to him along, and he had been told what we most wanted.

June 22nd Simoda Bay

Various other articles were brought in this morning from the shops in town, and trade was quite brisk, three or four shopmen having the privilege of displaying their wares on the boards. The idea that all this trade & negotiation and discussion had been carried on in a heathen temple, as if the Americans had come and shown their disregard of Japanese superstitions, and the little dread they had of all the idols of the country, by setting themselves down in one of the fanes, putting the gods behind the screnes in darkness & neglect—this idea sometimes came across me in singular juxtaposition to the actual proceedings. The Commodore sent some tea and glassware to the Commissioners, and arrangements were concluded about the rates of pilotage, the prices of wood & water, and some other matters. The stone for the Washington Monument came aboard, & by mistake the bill for getting it out was [~~made~~] [~~out~~] forwarded, from which we learned that the officials were expecting the moderate sum of [~~nearly~~] $80 for this single block, only a cube of 3 feet! They charged $72 for the two gravestones, and $32 for the fence around the yard, both of which rates showed their desire to make the best of our demands. The gravestones were neat pieces of work, & the inscriptions cut in gold style, so that we had nothing to complain of on that score; we made them take a reduction of $12 on both stones, as it was stated before making them, that the rate would be $30 or $25 each.

All official business being over, Morrow & I took a last walk up the valley, over the hill into the upper part of it, & around by the side of the river, walking 9 or 10 miles, and finding many old faces and acquaintances along the road, most of whom, especially at Hongo, seemed really pleased to see us. The country looked charmingly, the rice was mostly transplanted, and gave a beautiful green hue to the hillsides and terraces, the hills above were dressed in dark verdure; and altogether, we were constantly called on to admire the successive beauties of the scenery. We obtained fewer flowers than I expected, but the most of those near the paths had already blossomed, and a few berries had become ripe, among which were those of the paper-tree. It was the only walk I had taken since our abortive expedition to find the seven ri limit with Bent & Maury, and was all the pleasanter for its rarity. We got back to Simoda about sunset, which on this solstitial day was nearly eight o'clock, tired and gratified with the excursion. If there is anything which had rendered the expedition to Japan pleasant time to me, it is the walk in search of flowers, and the greater freedom of intercourse with the people thereby obtained; these have been taken, too, with an agreeable companion in Dr. Morrow, so that we have both been pleased with our rambles, with each other, and with the objects of our search. I shall always recollect them with him, and they form the pleasantest remembrances of Yokohama, Hakodadi, & Simoda; altho' elsewise I have nothing to complain of. It is sad to see how few are the sources of enjoyment, occupation, or instruction, which those around me have or find for themselves in such a spot as this, where the ordinary amusements and company found in seaports are wanting.

They scold the Japanese, the Commodore, the ship, the Expedition, and their own evil tempers are never blamed; truly, it is sad to see such perversity and waste of time.

June 23rd [Simoda Bay]

Soon after breakfast, all communication with the

shore was stopped, much to the disappointment of many. Mr. Bent & I were sent there with final messages, which gave me opportunities to do some errands for myself & others, & take a last look at Simoda. Many of the shopmen had articles arranged on their boards, having learned to exhibit them if they wished to sell them, and seemed rather disappointed at being told their customers were gone. I have found some pleasant people among these shop-people, and have been surprised to see how much the women do in the manangement of trade. I got a crowd at the door in a state of great merriement by ridiculing a dull fellow with a shrewd wife, for being forced to ask her opinion on the prices of things we wished to buy. In every shop, almost, a woman comes to the board, & in all she is present, for the family lives in the rear, which is not screened in any way from the ship or street. The custom of sleeping on the same mats, which by day have served for eating, gives more room in a house than with us, who set apart so much space for bedrooms. The loft, where there is one, seems to be more often used for storage than sleeping.

We returned aboard one o'clock, the steamers having gone out to the mouth of the harbor and made every preparation for an early start in the morning. The artists and others connected with the Commodore's suite, have all gone to "the Mississippi", printing-press, dogs, cats, barge-men, orderly, servants, boxes, birds, all, except Mr. Perry & myself, for whom there is no room, and Dr. Morrow, who is in the "Southampton". The Supply & Macedonian are to go to Killon to find the coal-mines, & then to visit Manila, chaplain Jones taking charge of the Macedonian. Mr. Boudinot goes aboard the Macedonian, and Mr. Mish back to the Mississippi.

In the afternoon, Yenoske came aboard the flag ship,

and brought off a number of parting presents, together with the birds & dogs for the President. He and Isaboro were in good spirits, and Com. Perry entertained them with cake & wine. He asked them a variety of questions too, one of which was about the results of the Phæton's raid in Nagasaki harbor in 1808. Moriyama said, that the governor, whose name he gave us, two of his colleagues(like Kurokawa & Ishia I suppose), & ten others, all committed suicide in consequence of the attack & detention of the Dutchmen. He said that all men of character avoided disgrace and capital punishment by suicide, ripping themselves across the belly & then cutting their throats; but that common people usually hung themselves. Regicides & murderers of superiors were transfixed with two spears and then decapitated, as they hung on a cross; common criminals were dispatched by decollation, but crucifixion or starving on a cross was not common. He said he should readily make way with himself, if he got into any trouble or disgrace, & the rest seemed not surprised at the assertion. When told that the Captain of the Phæton[39] was now admiral at Canton, and might be up in Japan next year, they were much startled, but were recommended not to dispatch themselves, but rather make friends with him, and drink his champagne. At leaving, the Commodore gave each of them a bottle, and they went away, shaking hands all round. They had gathered up all the Chinese cash we had paid them, and brought it back, preferring to return it at 1600 to the dollar, tho' they took most of it at 1200, rather than keep it.

I went with them to the Powhatan, where they paid over some more cash, and received some more presents. Moriyama & Isaboro gave me their names on a slip of fancy paper they had brought with them, from which it appears that the Japanese have the same custom of a 姓, a 名, a 字 and a 號 as the Chinese. The Siogoun, aged 44 now, is named Zhiun-na Goö-gaku Rio-in no Betto Genzhi no Chioja Ken Sadaizhin; the mikado is an older man, but the siogoun's name was so long, I did not ask for his superior's. Isaboro's name in full, is Genzhi Yoshimasa Tsu-shio Gohara Isabaro, 姓源氏名義適通稱合原猪三郎 the two first of which form[s] his surnames and all the rest his given name or names. His present official title is Kan Simoda Bugio kumi Noriki ohoshets Gakari, 官下田奉行组与力應接掛 and that of Kurokawa, his superior, Simoda Bugio Shi-hai Kumi gashira, 下田奉行支配组頭 that is, Imperially appointed to be assistant colleague to the head (officer) at Simoda. He is generally called Bugio or Bunyo or Bungio, the differences being caused by the sound of ng given by some persons & not by others. These officers are now appointed under Izawa and Take-noüchi, and expect to reside here permanently.

Our visitors took leave about dusk, and this closed all intercourse with the Japanese for the first American Expedition to Japan, being within three days of a year, by their reckoning, since it anchored off Uraga.

June 25th [*sic* / June 24th] [Simoda Bay]

A supplementary boat went ashore this morning from the Mississippi to carry some printed copies of the port regulations and rates of pilotage in Simoda, to leave with the authorities, so that the last visit was on our part, after all, as the first visit last year was on the side of the Japanese. The day began so rainy and the sea was so rough, we have lain at anchor all day, no communication being had with each other on the shore. I wish much to take another ramble over adjacent hills, but there was no chance; they appeared more inviting than ever, and at any time they and the country about this port are not excelled by any harbor we have been in in Japan.

On a review of the proceedings of this Expedition, no one can refuse his assent to the assertion that it has been peculiarly prospered by God, and so far as we are at liberty to say it, was planned & carried out so as to receive his blessing as a step in his plans for the extension of his kingdom in this land. The appointment of a naval man as the envoy was wise, as it secured unity of purpose in the diplomatic and executive chief, and probably Perry is the only man in our navy capable of holding both positions, which has been proved by the general prudence & decision of his proceedings since he anchored at Uraga last July. It has been favorable to his unbiased action, that he has had no captain under him, whose judgment and knowledge entitled them to the least weight in his mind; all, except Buchanan, spent their thoughts in criticising what he did, and wishing they were going home. If the Commodore & the Envoy had been two persons such a state of feeling in the officers might have at last crippled the firmest purposes of the latter, and thwarted the whole enterprise. But such a dilemma was avoided, and Perry regarded all under him as only means and agents to serve his purpose, perhaps too often disregarding wishes and opinions of a comparatively trifling nature. But that extreme is almost unavoidable in minds of strong fibre, & bred for years to command, as he has, such power has habit.

Further, the remarkable weather experienced since Perry left Macao for Shanghai last April—fair, pleasant & healthy in a degree to draw the attention of all, who have more frequently cried out, "See Perry's luck!" than been disposed to acknowledge the hand & favor of God in it—has not a little aided the Expedition. Four or five of the ships have grounded, but none have been injured; the Supply was ashore two days on the North Sand at Wusung, & thumped the rock in Simoda Bay, but apparently received no damage; the Powhatan narrowly escaped ruin near Labuan by striking a rock, losing only her fore foot; the Macedonian & Lexington grounded, but were soon relieved; and the Susquehanna got no damage by running on a bank in the Yangtsz Kiáng. The mistake made by the Susquehanna in coming to Yedo Bay, opening that of Sagami instead of Yedo, enabled the Commodore to tow off the Macedonian from her sand-bank before she received any injury, and to go up before the town of Uraga in imposing array; three powerful steamers like the Susquehanna, Powhatan & Mississippi carrying each another vessel, the Vandalia, Macedonian & Lexington, showed the Japanese the means we had at command, and may have inclined them to receive us now we had come and not refer to the strong letter they had written Perry through the Dutch, requesting him to stay away for three years. It seems to me that he who refuses to recognize the hand & blessing of God in these preservations, and involving his general approval, is unwilling to recognize it anywhere or in anything. The simultaneous arrival of the Saratoga & the steamers at Lewchew last year, & of the six ships at the mouth of the Bay of Yedo this year, prevented all delays; and so has the regular passage of the store-ships to China & back to Lewchew & Japan, to Hakodadi, to the Bonins and to Simoda from Kanagawa, carried out the plans depending on them. The long passage of the Saratoga last March is almost the only case of delay, and this caused no embarrassment. The general good health of the 1600 persons in the squadron, destitute as almost all of them have been of fresh provisions since last January, and the good condition of most [~~part~~] of the stores brought on, calls for particular mention, as the converse might have hampered the whole enterprise. The Japanese could not easily collect fresh provisions for so large a body of people, and the extremity of sickness might have driven us to the extremity of forcibly supplying ourselves with food at some rate, even if the alternative was instant hostilities and the attack of Yedo itself. Such a procedure, necessary as we might have deemed it for our own preservation, & not to be thought of in almost any position, might have been resorted to by some one less patient; and I can conceive might have removed the peaceful opening of Japan to an indefinite period. Now, not a shot has been fired, not a man wounded, not a piece of property destroyed, not a boat sunk, nor a Japanese to be found who is the worse, so far as we know, for the visit of the American Expedition.

Some will ask, what has been gained or done by this Expedition, at all commensurate with the cost it has been to the United States? What ultimate results will be seen must, indeed, be estimated, & can only be, when time has disclosed them, both in respect of trade between the two countries and intercourse between their people, in respect to the facilities Japanese coal can give to connecting California and Asia, and in that of supplying whalers and other vessels with provisions and retreat from storms. But in the higher benefits likely to flow to the Japanese by their introduction to the family of civilized nations thro' the Treaty of Kanagawa, and increased by the additional regulations signed at Simoda, I see a hundred-fold return for all the additional expense the American government has been at in sending out this expedition, and a mode of expending her income which will redound greatly to her credit. By permission of the Commodore, I drew up a paper of general character, which was sent to Lin last evening by Moriyama. In it, I endeavored to show how Japan could learn much which would be of enduring benefit to her by adopting the improvements of western lands, and allowing her people to visit them and see for themselves; adding that it was to set before them the most useful and curious specimens of western art that the President had sent out to them such things as a steam-engine, a telegraphic apparatus, a daguerreotype, all sorts of agricultural implements, books and drawings explaining these and other things, and not merely curious articles or eatables or arms, from which they might learn how to make such, or obtain the assistance of those who could instruct them. The great change in the policy of western nations, from what it was 200 years ago, was referred to as removing all grounds for fear of any evil consequences resulting to them by a greater extension of liberty now granted, and that no one could wish them to do aught which would be injurious or hazardous. The paper closed by a hint respecting the danger, if Americans were followed by spies and officials wherever they went, and that all that was necessary was to have those who did wrong, accused & properly punished.

Whatever results may ensue from this and many other hints given to the Japanese, since we reached the Bay of Yedo, I think that on the whole the impression left on the people of the squadron, has been favorable. More intimate acquaintance would show more good & evil traits in our character, and they have now probably seen a fair average. Erelong I hope & pray that the gracious designs of Providence in thus favoring this Expedition will be still further developed, and the light of reveal'd truth be permitted to shine upon the benighted and polluted minds of this people. The glorious promises, yet unfulfilled, of the days of gospel liberty, are evidences enough of what forms, at least, a part of God's plans in opening the way as has now been done. Among a people so inquisitive and acute, it cannot be long before some will be able to break away from the trammels which now bind them to Japan, and see, for as long as they wish, what Christianity has done for other lands, & what it will do for their own. The day of God's visitation will be one of love till the ignorant and degraded have had the paths of knowledge & purity laid open for them, and the page of Revelation put before them in their own tongue. In all this, I see a vast reward for the expenses of this Expedition, and a gain to the cause of humanity & goodness, beyond calculation in the paths, gold & silver or traffic.

In reviewing the proceedings of the last few months, it is fair to give the Japanese officers the credit of showing none of that hauteur and supercilious conduct, which the perusal of books might have reasonably led one to infer formed a part of their character. Compare the conduct of the Burmese when Crawford went to see them at Ava, or of the Chinese when Amherst went to Peking, with that of Hayashi and his collegues, and down, too, in the subordinate ranks of officials, a class, who are noted in China for their contemptuous treatment of foreigners, and every one must admit their superiority in point of courtesy, their decorum, their willingness to receive suggestions, and their general good sense in discussing the matters brought forward for their acceptance. Perhaps more impracticable men could easily have been found, and these seven were probably chosen for their views being favorable to a change in the national policy; but the other qualities referred to may fairly be taken as part of the national character, since we have seen them among all classes to some extent. In no country could more agreeable and kind-hearted men be found than old Yendo and Fuzhiwara at Hakodadi, and if one could conserve with all, he would find some traits to please him.

June 25th Simoda Bay

The whole squadron lay in the harbor yesterday, and we were forbidden to step foot ashore; tho' a ramble in the cool breeze blowing over the hills would have been most pleasant; not a Japanese boat came near us, and night closed over the harbor without any other communication there. Mr. Bent went ashore to take copies of the Regulations and Pilot charges, which had been printed for the Japanese in Dutch & English. This morning, the five ships got under weigh, but the wind dies away before the Macedonian & Supply could get an offing, & they had to anchor, although the former continued to get the assistance of several native boats. In this position of affairs, the steamers left them in the harbor, we taking the Southampton in tow, and soon Japan was lost to view. Doubtless our departure was a relief to the overburdened town of Simoda, for during the last few days, almost no provisions were to be procured, and yesterday morning we saw the long trains of Lin and his colleagues, walking along the beach homeward; Kakizaki on their return to Uraga & Yedo. After such an exit, the townsfolk would hardly recognize their own quiet village, if the procession of officials in Japan is as much a scourge to the common people as it is in China. There must have been a thousand people in the procession, and their various insignia found rather a picturesque train.

July 1st Lewchew

The passage hither was over a smooth and pleasant sea, the southwest monsoon being just strong enough to keep the ships well ventilated. On the way down, the Mississippi went near the island of Oh-sima, a large islet lying nearly a hundred miles north of Lewchew, to ascertain its size, and whether any harbors existed. Mr. Maury went ashore in a boat to reconnoitre, and as he approached the beach, was met by a party of natives, drawn up in arms, to oppose his landing. One among them had a matchlock, & one, who seemed to take the lead, had a single sword; others were furnished with stones, sticks or spears. Sam Patch soon undeceived them, [~~the~~] and stated the pacific intentions of the boat, when many of the men left, and got ashore, and some provisions were brought down to the beach. Mr. Maury slipped away into a village from whence the natives had issued, and found it a most miserable collection of huts, the abodes of filth, ignorance and heathenism. The men wore pins in their hair like the Lewchewans, while the presence of swords indicated their promiximity to Japan, with whose language their's had more affinity. They presented a more wretched condition, even, than any of those people whom we have yet seen, and cause one to notice how easily man deteriorates in a small community, where every member is compelled to labor for a living, so that there is no surplusage of produce on which a government can be supported, whose members, while they may oppress, still do much to maintain a higher state of civilization than the people under them do or would. These islanders, lying between Lewchew & Japan are worse off than either, and it is probable, because their little intercourse with either, leaves them ignorant of what is most worthy of imitation, and the feeble energies of their untutored minds prevent all efforts to better themselves. The shores of the island offered many patches of cultivated fields, probably of rice, and the hill-tops were mostly well-wooded; between them a few valleys opened, in which something like orchards appeared.

Yesterday, we spoke an English ship, the Great Britain, bound from Shanghai to England, from which we learned the news of the declaration of war against Russia by England & France, and some of the first steps in the dreadful drama. She first supposed us to be Russian steamers, and the officers who boarded her, found the captain & crew had been in a terrible fright, from which they had hardly recovered, tho' they had seen the American colors for nearly an hour.

On reaching the anchorage, Mr. Randall, Capt. Glasson & Mr. Betteleheim came off, to see the Commodore. The principal burden of their information was the murder of a seaman of the Lexington, named Board, on the 19th ult., & the injuries received by another named Scott, at the same time, in the market-place at Napa. Scott, & another comrade Smith, were buying something, for which they had paid the money, when an official took it away from the woman, at which they became angry, and began to drive him off. He called others, and Scott was soon thrown down, & so bruised as to be left senseless nearly. Both the sailors were at least tipsy, but Board would take nothing, and was not present when this attack was made, at least so far as they know, tho' he may have been coming up to their relief. Mr. Bubaur was informed at Tumai, that two of the sailors were lying in the street drunk, and as soon as he could went there, where he found the man Scott too drunk and bruised to help himself. While getting kago to take both of them to

Yumai, he was told that another was lying in the water near the causeway; and found the body of Board lying in a boat, wet and frothing at the mouth. The Lewchews said they had taken him out of the water, into which he had fallen & drowned. The corpse was removed to Dr. Bettleheim's house, and an examination by him & Dr. Nelson of the Lexington, showed that the skull had been almost broken by blows, & congestion of the blood on the brain followed; no spirit was found in the stomach, nor any flesh wounds or cuts on the body. The testimony of the Lewchewans was so contradictory that no reasonable account of the cause, provocation, or mode of death could be obtained; while his fellows were too tipsy to say what they did see or might have seen, if they really did see anything, and of course, we can get nothing satisfactory from them on the matter.

For some days after, the market was nearly deserted, & for more than a week, no one came to the house at Tumai. Mr. Bubaur had been stoned before this sad event; and Mr. Randall had written an earnest remonstrance to the Regent, which Mr. B., armed with a cutlas, carried to the castle at Shiu, or to that officer's house, & pounded away at the door, till the paper was received. A reply came next day, saying that it was a mistake, for the stones were not thrown at Mr. B., but the children had games of playing with stones, some of which fell near where he was passing! It was promised, however, that the children should be ordered not thus to play with stones anymore, but to reverently retire when they saw Mr. Bubaur. I wonder he did not inflict summary chastisement on them, when the deed was done.

The men left at Tumai have been supplied at stated times with enough to eat, & have spent their time in a quiet manner. The temperature has been generally pleasant, but the houses have leaked, for they are old and tiled.

Mr. & Mrs. Morton came in a little while after we had left last February, and have thus far received no molestation; they occupy the same rooms as Mr. Betteleheim did. Some letters were found awaiting us from China and U.S., which were too gladly opened by their owners.

In the day, Mr. Bent & I went twice to the Mayor's office, to make arrangements for a meeting with the Commodore & Regent, to demand the rendition of the murderer of Board, to ask for two stones for the Washington Monument, some flowers & birds of the country, the coins to be exchanged, & two pilots to go over to the Kirima Is. with a party of survey. A strange catalogue this, but likely to be followed by something as strange, and perhaps more instructive to these impertinent islanders.

During our absence the grandmother of the prince died, when the people went into mourning for 49 days, wearing no hair-pins, selling or killing no port or beef, and pretending to close government offices. The orders respecting flesh-meats was evaded by the people, and Mr. Bubaur one day came across the port market near the edge of a wood beyond Tumai; so that, it seems, here as well as in China, the people understand how much they are to value government edicts at in certain cases.

[Sunday], July 2nd Napa Roads

Dr. Bettelheim preached on board today, taking for his text, Ps. 118: 26, "Blessed be he that cometh in the name of the Lord;" in his discourse he drew a strange application & comparison of this text and the parallel quotation of it in Luke 19th, to our expedition to Japan. We went there on an errand of mercy as Christ went to Jerusalem; we went to Yedo, the seat of Japanese power, as Christ went to the Temple, the point of Jewish priestly power; he had a scourge of small cords, we had a powerful fleet, to drive out Japanese opposition, as he did the dove-sellers; he performed miracles to prove his mission, we gave steam-engine & telegraph to the Japanese, which might serve to them as miracles; we were sent by a great nation to knock at the doors of long secluded country, and open it to introduce them to the circle of nations, & Christ came to the lost sheep of Israel and of man to bring them back. There are comparisons enough to show what a strange jumble the preacher had concocted in his desire to welcome us back to Lewchew, & in his joy at the peaceable result of the mission. I hope never to hear another such. The Commodore had him & Mr. & Mrs. Morton, with others at dinner. Mr. M. seems to be more encouraged at the general look of matters here than he had reason to expect from what he had heard; and hopes soon to get hold of the language enough to use it.

July 3rd Napa Roads

I was sent for by the Commodore at five bells this morning, to draw up a paper respecting the murder of Wm. Board, in which he demanded a satisfactory examination of the criminals, and proper punishment of the guilty. He had proposed himself to go ashore, but concluded to send this document instead of Mr. Bent & 2 orderlies, and straitly intimate to the Regent that he would not be satisfied with any subterfuges. The paper was strongly worded, & when we arrived there, and refused to taste the provisions which were spread out for us, or to treat on any other subject, or to receive the birds and plants they had prepared in accordance to the request of Saturday, and also that no provisions would be accepted or bought until this serious matter was adjusted, and gave them then the document to peruse, the Regent began to see that we were in earnest. A long document was put into our hands, the same which had already been given to Capt. Glasson, in which, & in their reports, they adhered to the assertion that the man was drunk, &, after stumbling along as he went, had fallen into the water and was drowned. It seemed to produce no impression on them, to repeat & reiterate, again & again, that it was impossible for a man to fall so as to give himself such wounds in front & on the back of his head; nor could he rise up himself after receiving one of them, but would lie stunned. We remained till nearly noon, and left them to take the papers they had given us, to show Perry, refusing to touch a drop or accept a single thing. In the evening, we visited the two forts at the entrance of Junk river, to see their position, & then went by the spot on the causey where Board was picked up, around thro' the streets to the Mayor's office, where we found the Regent & officers still in waiting, & every dish remaining on the table, just as we left them six hours before. They all looked anxious, and when it was intimated that the Commodore was not satisfied with their reply, and gave them only till tomorrow noon to make suitable explanation & give the real criminals up for trial, they were still more perturbed and still; in fact, their silence was very impressive. The same story was repeated, but we would not hearken, nor taste a dish. Mr. Randall & Burbauer, with all the old sailors, are ordered on board ship, so that matters must look a little squally to these double-dealing people.

July 4th Napa Roads

Our passage and decided bearing last night had some effect on the Regent, for he and about a dozen attendants came on board the Mississippi this morning to see the Commodore respecting the case in hand, and get a respite of some days longer to examine some persons respecting the murder, amounting to some hundreds, then to a hundred, and then to a great many. As we knew well enough, from the papers already given in by them, that this examination of so many was a mere pretense, the Commodore very properly would not listen to their request for four or three days, nor even till tomorrow night, but on account of today being a holiday, he granted them till noon of tomorrow; & failing their rendition of the criminals, he threatened to take the possession of the forts at the mouth of the river & stop their boats. They asked for two days, but went away with this final answer, having first been shown some of the cobangs & ichibus obtained in Japan, the like of which they were expected to exchange for the coins we left with them, tho' Ichirazichi had the effrontery to assert he had never before seen them in Lewchew. It is probable that they are not common, but this was going rather too far; for if the Lewchews visiting Fuhchau have been known to have them, it is exceedingly improbable that one in his position has not even seen Japanese coins. However, his question, "If you have got them already from Japan, why do you now wish any more from us?" was a pertinent one, and I do not think Perry is right in pushing them so hard for coins which they do not make, when we know how stringent Japanese laws are on this point. The party left us in much despondency, and I do not pity them at all, since they have shown so much weakness & lying from the beginning as to take away all trust in their statements. For his homicide, they ought to receive a serious warning, which will leave those who come after us the safer, as well as Morton who is to live here. I am somewhat inclined to think the man Board may have been involved in a fracas with the Japanese crews there, & knocked into the water, where he was drowned, without any intention of killing him; & this still further embarrasses the Lewchewans, who like Balaam's ass, are between two walls; however, this is a supposition.

Fourth of July was kept by firing a salute of 17 guns from each steamer, by reading the Declaration of Independence, singing a song, music by the bands & the best dinners, which the larders afforded. The day was charming, and proved more of a holiday than Sabbath ever have usually been, so far as work was concerned, and in the moonlight evening, our ship's company was entertained by the singing of the minstrels.

July 5th Napa Roads

Work was resumed this morning early, coaling, making besides a court-martial on a drunken surgeon and the two sailors, who made the row in Napa. I was sent for from the Mississippi, and on going aboard found Ichirazichi & his colleagues, with a card from the Regent, requesting the Commodore to send some officers, and whoever else he pleased, to attend at the examination going on at the Napa kung-kwan. Mr. Bent & I went, and found the Regent & Chief Treasurer in the office, with two judges, sitting by the entrance opposite each other, & assistants or clerks on both sides of them, seven people, on the floor, two bailiffs below them, and still outside, on the ground beyond the porch, were two jailers, with a criminal or witness, between them, whom they were then examining. Heaps of ashes lay around the yard, and an awning or tent drawn back, was over the gateway, and a newly erected hut stood in one corner. Everything showed that we had finally set them really to work examining the case, & might now expect to get at the truth of the circumstances, so far as this deceitful people can speak it. After we had been seated a little while, the man, who was kneeling on the ground, his hands leaning on the porch, and uttering little more than repeated interjections of assent to the denunciations of the judges, was harshly seized by the jailor on his right, & his arms tightly pinioned behind him, and then each jailor gave him a heavy blow on his soles, a blow which might well nigh have broken the bones, had it not been so guaged that the end of the stick came down on the ground. However, rough as was this usage, the poor fellow gave forth no groan, nor moved his features, but repeated his responses of ho, ho, ho, to every interrogation or denunciation. As soon as he was led off by the bonds to the neat-shed, I called Ichirazichi, and told him that; as we could understand nothing of this examination conducted in the Lewchewan tongue, it was needless for us to remain any longer.

He replied, that they had been occupied since yesterday in reïnvestigating the case, and had not been able to bring it to a close, nor could they possibly do so before tomorrow night, for the number of people implicated as witnesses or actors was very great, and must all be examined. The authorities of Napa had returned an entirely false report upon the case, which the Regent & Treasurer there present had now ascertained. The facts elicited now, were, that Board had gone into a yard or house to trifle with or lay hold of a woman, who ran from him, calling out to a person in sight to assist her; he came in & seized Board round the body, who then struggled to escape, and got out into the street. Eight or ten natives had collected, who seeing the sailor pursued, & learning that he had attempted this woman, seized stones lying about the spot and threw them at him as he ran, hitting him on the head, and body. He fled for the water, and the populace, closing in as they heard the fracas, only made it more difficult for him to see any escape. Whither he jumped or fell into the water, or was pushed or thrown in, I did not learn, nor had the woman been examined.

This explanation of the causes and mode of Board's death was more likely than anything we had hitherto heard; but I upbraided him with the duplicity of the former report, its absurdity & imperfections, the supineness of the Regent in taking such a ridiculous report of a death, & not investigating it for three weeks, nor as soon as we had demanded the culprits last Saturday, and told him the day of grace was up, the time allowed had expired, and we must return to tell the Commodore. It was nothing to us what investigations they were making, for all we wanted was that the criminals be tried, & the authorities of Napa knew them already. It was the business of the Regent to see that the reports of subordinates were trustworthy, and if he palmed lies off on us we should hold him responsible. The life of an American was too serious a matter to be trifled with, however great was the provocation; & their nonsensical statement about the deceased having fallen into the water & nobody seeing it, made it difficult for us to believe anything they said.

The people around were as still as mice while we told them these things, and both the Regent and the fine-looking, remarkable old Treasurer, were so excited that they stood around the little table between us, hearing it all. I have hardly seen any person in my life present a more dignified appearance than this old man; his white beard reaching to his girdle, his gold pins [~~on~~] [~~the~~] in a hoary head, and his clean, flowing, whitish grass-cloth robes, altogether formed a beautiful picture. I wish he was more honest.

We left the draft of a treaty in their hands, consisting

of six broad articles, which Perry intends to get the Regent to sign, as a pact between the two nations. Some of its provisions extend over others, as well as all Americans. As we came off, another poor fellow was brought up for examination, & [~~pinioned~~] as the former one.

July 6th [Napa Roads]

The Commodore made no move yesterday afternoon, though I think it would have been well to have landed a party of mariners at the Ame-ku dera, to show that he was not inclined to longer delay, and when he set a limited time he meant to adhere it. However, it was not till after dinner today that he gave orders to Capt. Tansill to go ashore with 20 marines, and take possession of the temple & guard at Tumai, allowing no natives to enter or remain within the precincts.

After these orders were carried into effect, Mr. Bent went up to the Napa kung-kwa, where we found the Regent and another Treasurer in sitting, and the six judges & assistants, bailiffs, & all in order, as yesterday, but the jailors & witnesses absent. The awning was drawn over the yard, and more heaps of ashes were seen, indicating night sessions. All looked serious, but the Regent rose to receive us, and we told him our message, that some marines had landed at Tumai, & the Commodore wished him to go to Ameku-dera 天久寺 at 10 a.m. to meet him. The officers present had a long consultation among themselves, and then a list of six names were handed us, being persons who had been proved to have thrown stones, and were present in the mob, but it was difficult to ascertain whether these had hit the man, or who had instigated the mob. They implicated six others, who had not been examined, and therefore more time still was demanded to bring the case to a satisfactory close, but we refused to do so, as all the time they asked for had elapsed—that is, the shortest period they had stated.

I will give these islanders credit for much careful inquiry into this sad case; and we know that many poor fellows have been pinioned and pounded already in their injuries, and the chains lying around might tell more fearful stories, if they could speak. In a similar dilemma in China, it is more than probable that two or three wretches, guilty of some other offense, would have been brought & given over to us to do what we liked with them, and the officers would thus have [~~soon~~] washed their hands of the matter as soon as it assumed a serious aspect. Indisposed as I am to let the Lewchewans off for their outrage on Board, or to excuse their mendacity in the report palmed off on us at first, I am willing to do all justice to their present efforts to get at the real points of the case, and even to infer that a criminal here gets as fair an investigation as anywhere east of the Ganges. The system of espionage is so well established that it prevents many a crime by rendering its detection so easy; and the rulers can therefore afford to do honorably, in their view, when a case comes before them. Great cruelty is exercised, doubtless, in our view, but a criterion of that sort does not suit this latitude, any more than we ought to blame Bacon for his judicial cruelties as much as we do Jeffrys.

One of the judges was called up by the Regent, while we sat by, and as he respectfully stood slightly bowing before him, his white beard reaching to his girdle, his hair neatly done up & his clean grasscloth flowing dress, altogether gave him, in our opinion, as venerable and dignified an appearance as we had anywhere ever seen, far more so than anything we had met with in Japan. Mean and simple as this Lewchewan court-house is, such men as are here convened, to do what they seem (or feel) due to justice, raises one's opinion of the nation, and adds new respect for their institutions. And then too, whatever may be reality, either as to the provocation offered by Board to this woman, or her disregard or rejection of his offers or attempts, we certainly must place external morality in Napa greatly beyond what it is in Simoda, & Lewchewan officers above Japanese, for decency and respect.

July 7th Napa Roads

I was sent for soon after breakfast, and on reaching the Mississippi, found Ichirazichi and his cross-looker there, and judged by their countenances that they had some serious matter on their minds, which the suspence the delay had kept them in, had not diminished. The Regent had sent them off to propose a meeting on board ship to avoid the inconvenience to the Commodore of going ashore, but doubtless to save himself the mortification of visiting him at Ameku-dera, where armed men showed that he was no longer entire master of his beautiful island. The Commodore very courteously allowed the proposition, and Mr. Bent & I went ashore to tell him explicitly the terms on which he would be received. We found him and the Treasurer at the kung-kwan, and informed them that the Commodore was willing to meet him if he brought the principal criminal on board & gave him up unconditionally to him, and was ready to sign the treaty which had been proposed to them. They were not quite prepared to do this, and brought foward the Commodore's declaration that he did not wish to try the criminals himself; but I told them that I had said nothing about trying them, and as one American was killed, only one Lewchewan was demanded, and they need not bring off the six. After long consulation among themselves, in which most of the officials present joined, we left the office with this ultimatum, and that they would not be allowed to come on board otherwise, tho' they could not, as usual, be brought to say Yes.

At noon, they were alongside the ship, and the chief criminal with them, and were soon seated in the cabin, he kneeling pinioned before all. But the least hint had been given them of what was to be done with him; and when, after I had given Perry the purport of the proceedings, in which the circumstances of the rape were given as the provoking cause of the mob, and that this man had been found guilty, and been sentenced to bunishment for life to Pachung shan, and the other five to Ty-pin san for eight years, he replied, that he was now satisfied with proceedings of the authorities, and with the examination & finding they had made, & now gave the whole six back into their hands to be punished as they had decreed, their surprise and relief was so sudden, that the two chief, and all the other officials immediately rose up to make their profound acknowledgements. They perhaps thought the least punishment would be imprisonment and death, but the Commodore had it in mind to take him to America, whence he might be returned at some future day, qualified in some measure to benefit his countrymen. However, he told them he should leave the matter in their hands, taking their sealed declaration that the sentences had been properly executed. Respecting the articles of the Treaty, the Regent requested time to confer with the other Treasurers, & they would be ready to discuss the paper tomorrow and settle all its points. This was agreed to, and a meeting between the principals arranged for Monday. The Commodore also told them he wished a bell to hang in the top of the Monument at Washington; and I really believe he thought more of the procurement of this bell than the settlement of the case of murder & mob. The relief they had experienced led them to listen readily to the request for a bell, which belike will be used in the Monument to call people together to hear fourth of July orations.

Thus this difficult question has been satisfactorily settled, & in such a way too, as to leave an impression on the minds of the Lewchewans that the lives of foreigners are not to be trifled with; but that we, at least, are willing to do justly by them, and desirous to judge this matter fairly. This case was an aggravated one, and they are excusable, if any people could be; tho' to leave it with their merely making an apology would never do, and might be prejudicial to the safety of whalers or small vessels stopping here, if not to Mr. Moreton & his family. We of course cannot certainly tell what the authorities will do with the criminals, but I am inclined to think they will take a journey to the Madjico-sima.

July 8th Napa Roads

During the forenoon, the Commodore, who is as uneasy as a man with the toothache, and seems happiest, when stirring somebody up, was arranging and disarranging the presents he intended to send to the Lewchewan authorities, altering the lists, but never coming nearer to satisfy himself. A pailefull of beautiful fish, among them Spari, Balistes, Merra, and Aulostomus, brought in by Maury, offered a new subject for him for some time, until he got the artists at work painting them, calling them off from their dinner, lest it should not be done soon enough. The variety and gay colors of the fish in these waters exceed anything I ever saw before, but those we get are mostly from the reefs; and coral reefs are noted for gay fishes.

In the afternoon, we met the Regent and Chief Treasurer at the Napa hall, and now were happy to partake of their good cheer, which evidently afforded them satisfaction. The birds and plants were brought out again, one of the former being supplied with a platefull of musquito larvæ, wriggling in a little water; if birds were only able to feed themselves with these insects, Lewchew could support as great an aviary as any country I ever was in. The sojourn of Tansill & his marines for one night at Ameku-dera nearly used them up, such an attack did the musquitoes make on them.

At the meeting this afternoon, we discussed the various points of the treaty, they having carefully looked the document over. To our surprise, the greatest objection they made was to the preamble, in which it was stated that Lewchew and the United States entered into a treaty of amity, saying that this would offend the Chinese emperor, to whom they gave their allegiance, and who would visit upon them if they assumed an independent position, as this preamble asserted. In reference to Tuchara or Japan, they said that the trade with Satsuma was carried on mainly for the purpose of procuring rare & fine articles to carry with them to China when they took tribute to Peking. They wished to say nothing respecting this latter trade, and evaded a reply when I asked them if they did not take tribute to Kagosima also. The admission of being tributary to China seemed to please them, rather than be a humiliation, and the real fealty they are in to Satsuma must be a sore subject and a grievous burden, or it would hardly be so mortifying to them to say aught respecting it. Of course, if they are willing to promise all we want, it is likely to be held fully as binding to give the assurance in their own style. They tried, too, to get all the trade into the hands of officials by making it the duty of the captain of the ship to furnish a list of what he wanted, but this was refused, tho' we altered the clause which they so interpreted as to oblige them to buy as well as sell.

They defined illegal acts, for which all citizens of the U.S. can be seized and taken to their captain, as including "rushing or intruding into houses, ravishing women, forcing people to sell things to them at their price, and going about streets at night", from which I infer that these acts have been the chief obnoxious doings of Americans whilst here. We assented to this addition, except the last clause.

Finally, as Com. Perry had stipulated these liberties for all Americans, English, French, & other Western nations, they supposed he had authority on these points, and they wished to have him carry Mr. Moreton & family away when he left. As the inference was a fair conclusion from the premise, we did not reply, otherwise than promising to mention the matter to Perry; & such was their readiness to catch at even this slight but fallacious prospect, that both the Regent & Treasurer rose to return their profound thanks. This incident proves the wisdom of the Commodore last January, when he declined to give Moreston a passage in one of the ships. A sealed document was given to us by the Regent himself, containing the promise respecting the criminals:—

"A sealed declaration. A sailor of your country, named Board, on the 12th of June, about 4 o'clock p.m. forced his way into a house and violated a woman, and then rushed from the place; an angry crowd now came together, and some threw stones to wound him, others to drive him off, causing him to flee away, by which he was drowned. We have carefully investigated the case in all its circumstances, and adjudged to the criminals the following sentences, and have hereto affixed our seal as evidence. To the murderer Tokisi, at 29, 渡慶次, of Higasi-mura, for throwing stones & wounding the American, by which he fell in his haste into the water & was drowned, banishment for life to Pachung-san.

"To abettors in the murder, Konishi 國吉 æt. 16, of Komi-mura 久米村, Yara 屋良 æt. 18, of Watanji 渡地村, Arakaki 新嘉喜 æt.19, of Higashi-mura 東村, Chining, 知念 æt. 18, of Nishi mura, 西村, and to Karagusku, 金城 æt. 32, of the same village, banishment to Typingsan for 8 years.

["] Signed by Sho Fu-fing 尚宏勳 Superintendent of affairs in Lewchew, and Un Tukuyu, 翁德裕 Chief Treasurer. July 8th 1854."

The other two treasurers, Mo Fu-mi 毛鳳鳴 who came off to the Mississippi yesterday, & Ba Rio-se 馬良才 seem to have no jurisdiction in this case. Besides the above sentences, our friend the old Mayor of Napa, Mo Zhiukuring 毛玉麟 is deprived of pay but retained in office; and four sub-magistrates, Ri Yung-sho 李永昌 Zhiu Zaidén, 牛在田 Zhia Bunmo, 謝文茂 and Gu Fitsuching 呉必振 are all turned out of office; all for making a false report of the matter at first, which mislead the Regent. It would relieve the state of a great rascal, I think, if Ichirazichi, was sent off to the Majico-simah with the party, to stay there until he learned to speak the truth. Alas, he knows not the holy God of Truth, & why do I thus sit in judgment on him?

We gave the officials some other orders, adding an injunction respecting the bell, and the exchange of coins, by which time it was so late, that all wished the conference to end. We declined to take their version off to the Commodore, but waited for them to make a draft of the corrected copy. Thus Lewchew is likely to take, erelong, a more respectable position as a nation that she has hitherto done, and this compact will bring in, I trust, lasting good to these mild and peaceful islanders.

[Sunday], July 9th

Mr. Moreton preached a truly gospel sermon on board our ship today, wherein the great principles of religion were plainly made known, and gave much satisfaction to the audience. Dr. Betteleheim repeated his strange discourse, delivered on board this ship last Sabbath, before the Commodore, a stretch of metaphor and comparison which some did not hesitate to describe as blasphemous. I hardly thought he would have the assurance to utter such a sermon before Perry himself.

July 10th Napa Roads

As we landed this morning, the birds and plants presented [~~by~~] to the Commodore were going aboard, and when we reached the town-hall, there were the Regent & Treasurer, as if they had been sitting there since we left them on Saturday night. We discussed the various points of the Treaty, to most of which they agreed, but made more objection to the conclusion, desiring to have it read that as the Commodore ordered these various points, they humbly consented to allow them; but as this arrangement was inadmissable, they at last agreed to express it that they consented to it, he signing it first, and they affixing a seal only to authenticate it, and avouch their willingness. Fear of China was the only reason they assigned. It was a singular discussion; we desiring to have them sign this document on terms of equality as a sovereign state, & they debating every inch, preferring to own their subjection to China and great inferiority to us. They wished us too, to express instead of "western nations", the names of England & France, which we could not do, since that would offend them & be invidious to others, and therefore took it all, which made it unnecessary to say anything further concerning Mr. Moreton, about whose removal they gave us a long paper. Besides the discussion relating particularly to the treaty, there was some about the exchange of coins, which they still persisted in not having, about the size of the stones for the monument, and also relating to the bazaar, the whole interspersed and alternating with soups, melons, tea, cakes and other solids, served up to keep us in good spirits. They could take no more effectual way to get rid of us than to let us have whatever we asked for; it would act as well as it did when the Israelites went up out of Egypt.

These consulations were listened to with close attention by the bye-standers, but every one was agog when we opened the two lorgnettes & dressing case, to have a peep thro' them; and the treaty faded in comparison. In this nick of time, we told them the Commodore wanted a bell, a big bell, a bell as high as the table, a bell like the one at Ameku-dera, a bell which would make all ring again; and happily, a bell they straightway promised. It was at Shui, but could be sent for; truly, when it came off to the ship, it answered most of the stipulations, but it was cracked, and so was returned in the boat it came. I think they must have thought us cracked too, by the way we asked for this bell. If it ever gets to the top of the Monument, won't it utter Perry's glory or folly?

When we returned on board, Perry was passably satisfied with our report; & after dinner I slipped ashore for a stroll with Dr. Green, the first I had since Simoda's last.

July 11th Napa Roads

The various agricultural implements intended for the Lewchewans went ashore this morning, and all were arranged in good order in full time to present to the Regent. There was only time to prepare four copies of the Treaty in English and Chinese, and the rescript of the Commodore respecting the banishment of Tokisi, the criminal in re Board. This paper was sent to them in reply to their finding, and stated that the Commodore was satisfied with the final examination & decision of the Lewchewan courts, & with the unconditional surrendry of the chief criminal to him; he had given him back to them, with [~~the~~] reassurance that their promise would be carried into effect, as a warning to the people, who were in future not to seize men making a disturbance, or stone them, or beat them themselves, but were to apprehend them and give them to the authorities to be dealt with according to the decision of the captain & rulers. By this course of procedure, good feeling would be maintained.

At noon, the band & marines landed at Junk Harbor jetty, and marched in martial array up through the market to the main-street, & then down to the landing-place near Capstan Rock, affording an unexpected treat to the townsfolk and market women. At the landing, the Commodore met the body, & was escorted to the town-hall, where the Regent & Treasurer had made every preparation for receiving him in style, spreading an awning, setting out tables, and cleaning up the yard. What a doleful story would that yard and room tell, if they could speak out all the suffering & injustice done there by the authorities during the past week in the investigation made. But all is covered over & concealed from us, and perhaps it is well that it is so, for we could not help it, even if we knew it.

All parties being seated, the list of presents for the Regent & three Treasurers was presented. To the first, a revolver & flask of powder, engraving of the Washington Monument, & all the agricultural implements; the 1st Treasurer, a dressing case & engraving; the 2nd Do. & 3rd Do., each a lorgnette, & engraving; beside 15 po. cottons to the old woman aggrieved and assaulted. The copies of the Treaty were then signed by Perry & sealed by the Regent, each party taking two. This document is rather an important paper for this people, and will be much to bring them into fuller intercourse with their fellow-men, & show them the benefit of doing so.

The dinner was served up in usual Lewchewan style; first, the table was spread out with 10 or 12 small dishes, and then the warm viands brought on, fish in many forms, vegetables, custard, minced meats, kidneys, preparations of flour, and cakes, to the number of seventeen. We at last got thro' them, and managed to extract one laugh from the Regent by telling him that the Commodore would like to take his cook to America, and teach him in return for instructing in Lewchewan cookery, the mode of dressing some of our dishes; he seemed hugely pleased at this, and it was the principal event of the dinner. These islanders exceed the Japanese in cooking dishes suited to our taste, as well as in the variety and care of their feasts. They have, in such occasions, an advantage over their masters in wearing no long unmanageable swords, too, as well as sitting in chairs instead of on the floor.

We remained about three hours, partaking of all the dishes, & enjoying a cool breeze, and left them, they pleased that they had got the Commodore's promise to ask the Governor of Hong kong or England on his return there to send and remove Moreton from the island; and he more delighted at having got the big bell now at Betteleheim's house, tho' he had failed in obtaining any coins. The Regent, besides the bell, sent a pretty present off to the ship to Perry, of two bullocks, paper, pipes, cups, jars, cloth and other produce of the country. Altogether, this last interview with the officials was unusually agreeable to all present.

July 12th Napa Roads

The bell has rung the coins out of hearing, and I suspect the Commodore will now give them up as not to be procured. It was brought aboard safely this morning, and bandaged & wolded and canvassed & painted & boxed & strapped, as if it had been a mummy just disentombed & ready to fall to pieces. Won't there be a ringing of Perry's praises when this bell gets to the top of the Monument! However, as it has heretofore rung the orisons of idols, it is no desecration to it to be made to sound out the praises of men who are more than dumb idols.[40]

I have been all day at the kung-kwan in Napa, explaining the names & uses of the various agricultural implements, while the Lewchewans wrote them. There was a fine plow, a triangular harrow, a fanning-mill, a corn-cracker, a corn grinder to make indian meal, a cotton-gin, a double yoke, various rakes, forks, shovels, spades, &c. Among them was a churn, which I asked the Lewchewans to tell me what it was; and after looking at it a long time, & considering that as it stood next to the fanning-mill it had some affinity with that, they concluded that it was a machine to place sideways and fan people as they dined. It might as well have been so explained, as for any use it will be to them as a churn. Most of the others were understood, and perhaps, some of them will come into use here, but so expensive are most of them as to be beyond the reach of this people, and others are too complicated for them to use for a long time to come. The cotton-gin will be thrown away, and had better given to the Chinese.

In the afternoon, various articles came in for the bazaar, much the same as were exhibited last year, but rather better and more in quantity. The dollar here is reckoned at 1440 cash, but all things are in proportion to that valuation, so we are served fairly.

[a Chinese document, dated July 11th 1854, 2 pages: the following is its translation into English] Articles of Agreement

Ⅰ. Hereafter, whenever citizens of the United States come to Lewchew, they shall be treated with great courtesy & friendship. Whatever articles these persons ask for, whether from the officers or people, which the country can furnish, shall be sold to them, nor shall the authorities interpose any prohibitory regulations to the people selling; & whatever either party may wish to buy shall be exchanged at resonable prices.

Ⅱ. Whenever ships of the United States shall come into any harbor in Lewchew, they shall be supplied with wood & water, but if they wish to get other articles, they shall be purchaseable only at Napa.

Ⅲ. If ships of the United States are wrecked on Great Lewchew, or on any of the islands under the jurisdiction of the royal government of Lewchew, the local authorities shall despatch persons to assist in saving life & property, & preserve what can be brought ashore till the ships of that nation shall come to take away all that may have been saved; & the expenses incurred in rescuing these unfortunate persons shall be refunded by the nation they belong to.

Ⅳ. Whenever persons from ships of the United States shall come ashore in Lewchew, they shall be at liberty to ramble where they please without hindrance, or having officials sent to follow them, or to spy what they do; but if they violently go into houses, or trifle with women, or force people to sell them things, or do other such like illegal acts, they shall be arrested by the local officers, but not maltreated, & shall be reported to the captain of the ship to which they belong, for punishment of him.

Ⅴ. At Tumai is a burial ground for the citizens of the United States, where their graves & tombs shall not be molested.

Ⅵ. The government of Lewchew shall appoint skillful pilots, who shall be on the lookout for ships appearing off the island; & if one is seen coming towards [~~the~~] [~~island~~] Napa, they shall go out in good boats, beyond the reefs, to conduct her in to a secure anchorage, for which service the captain shall pay the pilot five dollars; & the same for going out of the harbor beyond the reefs.

Ⅶ. Whenever ships anchor at Napa, the officers shall furnish them with wood at the rate of 3600 copper cash per 1000 catties; & with water at the rate of 600 copper cash (43 cents) per 1000 catties, or for six barrels full, each comtaining 30 American gallons.

Signed in the English & Chinese languages by Commodore Mathew C. Perry, Commander-in-chief of the U.S. Naval forces, in the East India, China, & Japan Seas, & Special Envoy to Japan, for the United States; and by Sho Fu-fing, Superintendant of Affairs (Tsu-li-kwan) in Lewchew, and Ba Rio-si, Treasurer of Lewchew at Shui, for the government of Lewchew; & copies exchanged this 11th day of July, 1854, or the regin Hien-fung, 4th year, 6th moon, 17th day, at the Town-hall of Napa.

(Signed) M.C. Perry

(L.S. of the kingdom of Lewchew.)

In respect to this agreement, whatever it may lack, it contains enough to bind the Lewchewans down to a regard for their fellowmen, and to treating them better than they have heretofore felt obliged to do, which erelong will do them great good.

July 13th Napa Roads

In the morning, Mr. Spieden & two or three others of us landed near Capstan Rock to take Mr. Moreton the amount ($275) subscribed for the benefit of the mission here. We found Dr. Bettleheim just going afloat with a boatful of baggage, including chairs, tables, and many things which surprised us in one going where such articles of furniture are plenty; & on reaching the house, we saw it was bare enough. Mr. Moreton merely remarked in reply to our observation, that he thought Dr. B. would have taken the house too, if he could have done so. Something must be wrong about Betteleheim to act in such strange ways; & when we heard how he had claimed half the money given to the mission, & had gone to Edgarton & some other sailors to ask them to whom they supposed they had given their subscriptions, his mercenary spirit was too plain.

I was occupied all day at the bazaar, where some $100's worth was sold, principally of common articles; the assortment was better, far, than last year. The traders committed the whole management to my hands, receiving my accounts of sales without even examining them. We have seen so much better things at Simoda, that these look very ordinary.

July 14th

Everybody remembered that one year had elapsed since the stirring day when we landed at Gorihama (perhaps more properly called Kuri-kama 久里濱) in such martial array, and when the Japanese made such efforts to be prepared for any treachery on our part, as we did also on theirs. Now the trial is made.

The bazaar was continued till about noon, when all the articles were carried off, and erelong the Regent and two treasurers came in to have their daguerreotypes taken; Mr. Brown did as well as the glare of the sun & their pertinacity in keeping on their light dresses, would allow. They utterly refused to go to Moreton's house, for by thus doing they would measurably have acknowledged his existence. Soon after five o'clock Mr. Draper came in to let them know that the boat was ready. The Regent got into his chair or kago, borne of four, & squatted down at his ease. In the street, his retinue marched in front of him, spreading as wide as the street; first, went two [~~yellow~~] [~~flags~~] men, carrying each a wai buchi, bastinado, made of the lower end of a large bamboo, tapering almost to point, and split rather smaller than the middle, both sides painted red, & in most respects like those used among the Chinese. Next these flagellants (for to punish evil-doers is their office) came two gong carriers, who gave their instruments two raps in unison; next, two flags, each marked 金鼓 kin-lú, or golden drum; and just before the kago, in stately pace, stalked two young pages or secretaries, and between them & the flags, were borne two balls of cock's tail-feathers at the end of poles twelve feet high; what these omoi signified, I did not learn. Behind the kago, went a boy with a camp-stool, two bearing each a waku, or open frame holding a tent, awning, or something of that sort. The cap-box & pipe, boys came last. Such is the dignity of a Lewchewan grandee, and while he passed, we two were the only persons upright, except the retinue itself. The Treasurers had flags but no gongs.

When they all reached the boat, it was curious to see how these attendants contrived to get into the same one with their masters, but except a few in the bow, we stowed them into native craft, & were soon alongside. In the evening, there was an entertainment of singing & dancing, with a burlesque of a row in a barber's shop, by Ethiopean minstrels, which amused them very much, notwithstanding their constant grave faces. This people, from high to low, put on an air of seriousness, & there is less merriment in the thorofares, than any place I ever visited. However, when the darkies tumbled over each other, & scattered the flour [~~over~~] [~~each~~] [~~other~~] about, even these quakers could not contain themselves. The diversion passed off very well, the evening was calm, & all the natives were ashore by 10 o'clock, evidently much amused. Dr. Betteleheim thinks it will furnish talk for the next two years.

July 15th Napa Roads

Early on shore today to settle accounts with the authorities, so that there shall be nothing to do tomorrow. They have learned how to charge pretty well, and I hope that the real owners of the provisions, and laborers, too, are beginning to receive some portion of what is paid; we saw, a few days ago, that when the men received $5 for provisions delivered in this ship, they paid over one to the officer in the boat. In settling up for the expenses of taking the coal off to the ship, the Lewchewans estimated 1017 days' work done in the 8 days it required to clean the coal-shed, while at a large average there were only 45 or 50 laborers actually engaged on shore & in the lighters, a new gang being sent to the shed each day. It appeared, therefore that the pay one official overseer received a day, was equal to ten to twelve common men, there being about eight drivers to urge up the tardy. In this proportion, two poor laborers take three officials to look after them. Their bill of $129 we reduced to $100, and that of $41 we cut down to $12, since as it cost only $58 to build the whole shed at first, $12 was plenty for thatching two wings and mending two ends. The Regent was admonished to keep it in order, and a flag was given him to hoist at the dépöt whenever an American ship came into the harbor, as well as a small one to take off to ships in the Roads when the pilot goes to conduct them in. How unlike this to the ignorance of the Lewchewans when the Morrison's flag was unknown, they having never before seen an American flag! In return for the two flags, the interpreter gave me a drawing of the Lewchewan flag, called 巴, & drawn like the trine powers' diagram. He said it was always hoisted by their junks going up to Fuhchau. The coat of arms of Kurokawa is precisely like it.

Some pieces of bullion were exchanged today for the coins left at the palace at Shui last February, but as they were useless as coins, they were all sent back, except 200 Japanese cash; and so the long contested matter was settled, & the Lewchewans carried their point. The two stones were also taken on board this morning, and one of them broken up for holystones, it being utterly unfit & worthless.

I was told today that the late Regent, Sho Rai-mo 尚大

謨 whose removal from office caused so much speculation last year, when we returned from the Bonin Is., was still living in Shui; he had resigned his position as Tsu-li-kwan from age, conscious of his inability to undergo the fatigues likely to come upon him tho' the squadron, and management of all its demands. No coercion was used, it was a voluntary resignation. This removes all the reports we heard then, and from the way I was told I am inclined to believe it to be true.

It appears that the present & last Regent are both allied to the royal family, whose surname is Sho, and they are cousins. The prince is now eleven years old, and will probably receive his investiture from China in four years; his name is 泰 Sho Tai. His father died in 1847 aged 38, leaving this son; his name was Sho Iku 尚育, & he had reigned about 10 years. The prince's grandmother, who died a few days after we went to Japan, was the wife of the king regnant when the Alceste was here in 1817; she it was who had been so alarmed when Capt. Shadwell went up to Shui, that she had been taking broths for 17 months, when we visited the palace in June last year. This palace is an extensive structure, much larger and exhibiting more skill than anything we saw in Japan. It was partly rebuilt, & thoroughly repaired about 20 years ago, but the wood work is rapidly decaying from the climate, no paint being used upon it, nor anywhere else in Lewchew. Its general design so much resembles a fortress, that one can hardly avoid concluding that such was one of the objects in view building it.

July 16th Napa Roads

Mr. Moreton preached in the Mississippi today, and Dr. Bettelheim in the Powhatan. The former remained with his wife to dinner; the latter has not been ashore since he came off with his baggage three days ago, and the coldness between them has attracted general animadversion, mostly taking sides with Moreton. The thanks he sent to the squadron for the donation was read to the crew of the Mississippi today, & did him credit. I pray God to protect & bless him in his loneliness, and preserve him from unreasonable men, who have no faith. I accompanied the party who landed him in the evening, after all communication with shore had been forbidden, and left him & his wife in their new home. The boat's crew left $4 for their son Philip, as they were shoving off — a handsome thing.

The daguerreotypes of the Regent & two Treasurers were sent them today with a portrait of Perry's, as a parting token of good-will. They were doubtless pleased to get them, as well as Ichirazichi, tho' none were superior. With this, closed the visit of the American squadron to Lewchew, but not its effects, nor I hope, its good effects.

The Lewchewan authorities, having learned that their old trouble, Dr. Betteleheim is leaving, are desirous to get rid of their new one, Mr. Moreton, and gave the Commodore a long paper yesterday, reiterating what they had told him before in respect to both the missionaries. It was written in the names of the Regent & Treasurer, Sho Fu-fing & Ba Rio-si, who say, "We earnestly intreat your excellency to condescend to regard us with kindness and greatly strengthen our affairs, by taking away to their own country, Moreton, who remains loitering here, in so doing, compassionating our little kingdom. It is well known that we are a trifling unimportant state, a country of no value, whose soil is poor & unproductive, as are likewise all the little islands dependant on it. Not only have they no gold, silver, copper or iron, but no silk, satin, or pongee; & so few and meagre are the productions that it is undeserving even of the name or style of a kingdom. Since the days of the Ming dynasty, we have been regarded as an outer dependency of the Middle Kingdom, from whose favor we have for ages received investiture for our king, and to which in return we have given tribute. Whenever there has been any important event in our borders, it has been reported; whenever the time came around, for us to send up the tribute, we have then purchased raw silk and goods to make up into dresses & caps for our various officials, and such medicines and other articles were selected as were necessary for the use of the state. If we were not able to procure enough in this way, we have exchanged our products, as black sugar, spirits, grasscloth, &c., with the island of Tanega-sima and friendly neighboring country, where we get things suitable for tribute, & send them to China.

["] Such things as are indispensable to us, as rice, grain, iron utensils, cotton, tea, tobacco, vegetable oil, machines, and other articles, are sought for in this island, whereby our necessities are supplied. Yet if the crop of grain here is deficient, people are forced to satisfy their hunger by sweet potatoes, since there is not a peck or a gill laid by in the country; and in times of storms or drought when the harvest is blighted, lamentable indeed is our condition, for we have nothing to eat, and as a substitute to prepare something from the iron tree (or Cycas) to save ourselves from starvation; or borrow corn from this island to supply our need.

"Our traders in the markets have only for sale tea, tobacco, wax, grass, shoes, melons, greens, cotton or grass cloth, old clothes, and other trifling articles of daily use, and this trafic is managed by women, being therefore utterly beneath the notice or glance of other nations. Consequently, when ships from western countries have, during the last few years, often come here, the various articles of daily use they have required, (what an assortment they were!) could not be procured in the public markets; we have called the officials & people to Napa, and sent some abroad to places to buy them, or taken other articles out of the public stores, which was reducing the stock laid up for the use of the state, and also hazarding a dearth in the returns of the farmers, both of which was dangerous and troublesome. In the years 1844 & 1846, some French officers came, and the Englishman Betteleheim brought his wife & children, to dwell here, all of whom needed supplies to be provided, difficult as it was for us to get them. Whenever ships of these nations came in, we have made known these circumstances to them, earnestly begging them to take away these persons. The Frenchmen, knowing the sad condition of our country, went back to their own in 1848, & have not hitherto returned here. But Betteleheim has been loistering here ever since, and has just now brought Moreton with his family to dwell in his stead, so that our people have no rest, our impoverished land no relief.

"Learning lately that your Excellency has control over the ships of all western nations in the East Indian, China, & Japan seas, and that none of them can go here & there to other countries without your orders, we have thus minutely stated our unhappy condition, and humbly look up to your abounding kindness, intreating that when your fine ships leave, you will take Moreton with you away back to his own land. Then will rulers & people be lifted up, & all will feel the effects of your great kindness, & wish you the happiness of seeing a thousand atumns."

Whether the Lewchewans will do anything to Mr.

Moreton to rid themselves of him, I think very unlikely; indeed, I rather think these repeated applications are urged by their Japanese rulers, who may change a little on hearing [~~of~~] what has been done there. The mission certainly has great difficulties in the passive resistance the people offer, and needs the Arm of its Almighty Protector to guide & shield it.

July 20th [Ningpo]

We left the harbor betimes on Monday last, being my sixth departure from Lewchew, and accompanied the Commodore till about 11 o'clock, when he took his leave, & left us to go on our way to Ningpo. The Captain took his course N.W. towards Video I., which was made yesterday morning, and a clear day enabled him to get down to the anchorage off Kintáng, below Lukong, before sunset. The day was intensely hot, increased as it was by the great fires we carried in our furnaces, and everybody was glad to see the sun disappear.

Two boats left the ship at sunrise this morning, and aided by a strong current, soon entered the Yung R. & stopped at a custom-house landing at Chinhai. The tide was so far spent, however, that no boats could be got of a suitable size to take us up to Ningpo against the tide, and nothing remained but to pull the twelve miles before us. A tedious, burning, pull it was, and the sun had passed meridian before we reached Mr. Rankin's house[41], almost exhausted with the sweltering heat & glare: therm. 97°. After seven months of sojourn on shipboard, it was very enheartening to be once more in the cheerful company of one's countrymen, and join in praise & prayer. We found the missionaries at Ningpo all well, two invalids, Dr. &

Mrs. Mcgowan[42] having gone to Chusan to recruit. We had, as we soon learned, come at most opportune time, not less to the surprise than the joy of our friends, for only a few days had elapsed since they had been placed in considerable danger by the violent proceedings of Capt. Lopez commanding the Portuguese corvette Don Joaō I., then lying off the Consul's. The circumstances are briefly these.

For some years the Portuguese lorchas have carried on a thriving business in convoying Chinese junks up & down the coast, in which they have committed so many atrocious acts against their customers, as well as the people along the coast generally, that they are losing it, and the Canton junks refuse to take their protection. The Portuguese stigimatize these men as pirates, and have had a number of collisions with them & their vessels, in which lives have been lost on both sides. This has created bad feeling, and the Portuguese consul Marquis, finding that his cause was losing ground, sent to Macao for the corvette. She came up, and the Canton men began to prepare for resistence. Things went on from bad to worse, the Consul & Captain thinking themselves invincible till the latter in an evil hour, took his bargue into the north or Tsz'ké branch of the river, nearly abreast of the houses of the American missionaries, and off the line of Canton junks on the other side of the river under the city walls. On the 10th he opened a fire upon them, having given no foreigners any notice of his design, and sure that many of his balls would go into the city, while, if the junks returned his fire, their balls would fly here & there among the houses of the Americans, putting them in imminent danger. However, the Chinese left their boats, & left without much injury, as did also our countrymen; but many balls went from the corvet into the city, injuring dwellings

& destroying five or six people. In one case, an old man was hit, & his son, walking on, heard that he was wounded & went back to assist him, when a second ball killed them both. The people of Ningpo were naturally terrified at these proceedings, and began to pack up their valuables & clear out; while the Tautai was totally at a loss what course to take. He had a conference with Mr. Meadows[43] & Dr. McCartee[44], and wanted them to promise that an English or American steamer should come down, which of course neither of them could do. No one could tell what a boasting Portuguese captain might do in such circumstances, and this posture of affairs rendered the Powhatan's arrival such a matter of congratulation to all, especially to Mrs. Rankin, as her husband showed us a ball or slug which had hit their house. So unprepared were they all, [~~however~~] for our appearance, that it was sometimes before they could be assured that it was not the Susquehanna. A letter was drawn up by Lieut. Pegram in the course of the day to send to Capt. Lopez, but as we were told that a conference was to take place on the morrow between him & the Tautai, he decided to submit it first to Capt. McCluney, pending the result of this interview.

Such was the hap we found at Ningpo. Our company was distributed around, Mr. Perry & the surgeon going to McCartee's house, Nicholson & the purser to Way's[45], Capt. Jones to Martin's, Mr. Randall & King to Goddard's[46], & Mr. Pegram & I to Rankin's, Cobbold[47] taking Bettelheim. After dinner we took a walk thro' the town with McCartee; and, at last, after 21 years in China, I have this day been inside of one of her cities. The Doctor was greeted by many persons, and we went thro' various streets & into many shops, everywhere finding a pleasant reception. The walk was prolonged until darkness overtook us, and we were glad to get out of the hot streets into the cool breeze on the river and the cooler verandahs of the houses. I found the streets of Ningpo more dilapidated, the houses less substantial, and shops, stalls and markets generally less extensive & bustling than I had expected, but probably much of the dullness was owing to the late commotion, and something to the time of day. The pái-lan were, many of them beautiful structures, and if there was more space around them to set them off, they would equal in effect many of the porticoes & pillars of European cities.

July 22nd Off Chinhai

Yesterday, about noon, Capt. McCluney sent off the launch, containing twelve marines with a howitzer and some ammunition, and his instructions to Lieut. Pegram to remain in Ningpo until Capt. Lopez gave the most satisfactory assurances that American lives & property should not again be jeoparded by his proceedings. The boat reached town about sunset, and there was some stir in the heretofore quiet premises of McCartee, as the marines marched into his yard, and the sailors drew brass fieldpiece over the pavement. There were about 80 persons now about the mission-houses from the ship, all of whom were soon accommodated with as comfortable sleeping-places as could be wished. The only thing mortifying to us in the eyes of the Chinese about the houses was the drunken conduct of a few of the sailors.

This morning, Lieut. Nicholson took the letter to Captain Lopez, who promised an answer as soon as he had conferred with the Consul. In the meantime, nothing could be done, and we hoped he would soon prepare one, for it was desirable to get the men again on board ship out of the sun, of which they seemed to have not the least dread. Near moontide, one of them was struck dead, falling like a log on the side of the path, & hardly conscious of any ailment or pain before life was gone. He was alone as he fell, but some of his comrades came up in a few minutes, and carried him into Dr. McCartee's dispensary. There was nothing to do for him, but give him a decent burial, which was done about sunset. Thus quickly was this poor man called to leave this world; he had drunk but little during the morning, tho' he was notorious for violent conduct when in liquor, & had already given trouble by going into a shop at Simoda, where he broke open saki pots and wounded a Japanese who tried to prevent his violent proceedings. Alas, for James Clark!

Having made all my purchases, I started at 2 p.m. with Mr. & Mrs. Rankin & others to visit the ship. It had excited great curiosity in Ningpo among the Chinese, & Mr. Martin took his school down in order to show the boys the wonder they had been talking of. Favored by the tide, we reached the Powhatan in less than three hours, and greatly gratified the officers, who were delighted to see three of their countrywomen, & after a cruise in Japan of seven months, to be so unexpectedly visited. Capt. McCluney did everything to make the short visit agreeable, and all were gratified. The party left about $7^1/_2$ o'clock, and reached Ningpo in upwards of these hours, both boats going back together. The ladies were much interested, & their coming aboard was much appreciated by the Captain, and amply repaid the trouble they had taken. If the ship had stayed at the anchorage longer, some of the Chinese officers would probably have visited her, and learned more therefrom than by reading a hundred books.

[Sunday], July 23rd Off Chinhai

It was well for me that I decided to remain on board last evening, for by morning all the boats had returned aboard, leaving one man at Ningpo, who had deserted. The captain of the corvette had sent in a letter, which was deemed satisfactory, and is likely to prevent his doing anything more which will endanger the lives & property of the Americans living at Ningpo. Capt. McCluney's intention was to force him to respect both, if he hesitated the least, for his conduct had been such as put him without the limits of all respect, & treat him like a brigand.

The men & officers were all so wearied & fagged by their long row of 14 miles in the boats that there has been no service today: the bum-boats alongside were bartering & chaffering their articles, however, till nearly sunset, officers & crew alike careless of its being a day of rest, and forgetful of the sudden death of one of the men, who only a day before had been in as good health as they were. Even with the best intentions, it is almost impossible for one to withstand such a current of vice, when it is compressed within a ship's sides.

Towards evening, Mr. Randall stopped at the ship on his way to Shanghai, and told us that Roach, the man who deserted yesterday had not been found, nor did the Consul know where to look further for him. So, our ship's company is reduced two men by the visit to Ningpo.

It is gratifying to me to hear the general expression of thanks among the officers who went up to town for the kindness with which they were received, and their opinion of the members of the mission. Living, as they do close together, it is easier for a stranger to visit the several families, and thereby obtain a definite idea of the size and arrangements of the mission, than when they are scattered over the city, as in Canton; perhaps there is an advantage in concentrating influence by associating it in the minds of the people with locality, that may not have been duly considered in other cities, when selecting a mission-house & station, McMartee is well fitted for his post as a consul, and his knowledge of the people renders him a good adviser to whoever may be appointed to hold that office; the less that missionaries are known as officials, however, the better for their reputation.

July 26th Fuhchau

We left Chinkai on Monday morning, and had a clear day to run down through the islands along the coast, and about noon today anchored off the White Dog Is., beyond which Capt. McCluney does not mean to take the ship, tho' merchantmen drawing more feet than we do run up the river within 12 miles of town, & we lie off forty from it. This morning we saw the Lady Mary Wood passing down the coast. The weather has been so threatening that Mr. Glisson declines starting, tho' a fair wind would carry him directly into the river.

Aug. 1st

Yesterday, Mr. Glisson, Mr. Perry, Dr. Schriver & Dr. Bettelheim, returned from town, bringing with them Mr. Tyers, who wishes to get a passage to Hongkong. Matters are going on smoothly at Fuhchau, and the disturbances at Canton seem to drive & draw more teas to this port, than have been brought hither before. The "Saml. Russell" is nearly loaded. Of the mission, Mrs. Cummings[48] is the only one ailing, & her case is not such as to excite alarm. Mr. & Mrs. Baldwin[49] are up at the Monastery. There have been some disturbances, but not such as to cause fears of an outbreak. This city may in fact be one of the last places afflicted by the insurrection which has carried trouble & ruin to so many other parts of China.

Aug. 5th Amoy

I came aboard this evening with Mr. John Stronach[50], who wished to see Dr. Young & Mr. Burns before they left Amoy. We have been here five days, and I have enjoyed the visit much at the houses of friends residing here, most of whom I found in good health, and all of them greatly encouraged in the work of God among the Chinese. Since the departure of the insurgents, the utmost quiet has pervaded the town, and the citizens are content with their present rulers as being less of a scourge than the new claimants for power, who fleeced & murdered wherever they could find prey, until the whole town groaned with their rule of iron. Business is gradually resuming its course, and in China this of itself indicates the return of peace and security.

The work of grace at Peh-shwui-ying or White-Water Encampment, a village west of the city about 12 miles, is one of the most encouraging things yet occuring in missions in all China. The work began last January with the visit of Mr. Burns[51] & two native assistants to the village, where they have remained ever since making known the glad tidings to inquiring natives, about 20 of whom have believed to the salvation of their souls. No reasonable doubt can remain in the minds of any one, acquainted with the facts of the case, as to the sincerity of their professions, and their zeal for the cause they have espoused exhibits itself in efforts to make its blessed truths further known. The number of members in Mr. Doty's[52] church is 52, and there are nearly as many more in other churches.

I visited the tombs of several missionaries on Kuláng-see, most of them women. Mrs. Boone[53], Mrs. Doty[54], Mrs. Pohlman[55], Mrs. J. Young[56], are among those whose graves witness their desire to tell the Chinese of a Savior's love. Mr. Lloyd[57] also lies here, & several children, and two converts; James King's grave is just opposite. It is altogether a pretty place, aside from the interest attached to it as the resting place of friends. I could not, for myself, have passed by Amoy without making a visit to the grave of my dear friend, Mrs. Boone, whose virtues endeared her to every one who knew her.

The city of Amoy presents a most singular appearance of dilapidation from the looseness of the tiling, which a breeze of unusual severity blows off or throws away. Just now, it is increased by the ravages of the late occupation, and the cannonading which both parties inflicted on various parts, some entire blocks being demolished. On Kulang-su we went thro' one village, whose empty streets, roofless houses, demolished gateways, and uptorn gardens, presented a melancholy spectacle of the lesser evils of war and strife. The common dirt in the streets of Amoy is bad beyond most cities, & abundant, but its noisomeness and depth increased while we were here by alternate rain & sunshine, which caused such odors to ascend and pools of filth to form, in the streets, as are not, I hope, excelled anywhere even in China. I do not see how people can live in the midst of such exhalations and in the confined, dark holes they were seen in. The many pale, sickly faces in the doorways proved that malaria had some effect. Few shops of any pretensions were seen, but still a pleasant activity and thrift were apparent, quite unlike the torpor of Ningpo. A new branch has lately sprung up, that of making artificial flowers, and the beauty and accuracy of the imitations must soon recommend them to foreign parts.

I took dinner last evening with Harry Parkes[58], and spent an hour very pleasantly there, Capt. Vansittart & some of his officers forming the rest of the circle. I was glad to see that consul was so much respected in the place.

[Sunday], Aug. 6th

We sailed this morning, having Mr. Burns & Dr. Young[59] & child for passengers to Hongkong. The insanity of the Doctor is much like that of James Bridgman, and he has tried twice to jump out of the windows, dreaming himself abandoned of God, & useless to the world. Today he is much more calm, & the new situation he finds himself in, restrains him from any repetition of the attempt. Mr. Burns gave the crew a gospel discourse, to which they listened with respect.

Aug. 11th Canton

In seven months from the day I left, I am permitted to return to this city in health. The steamer reached Hongkong in 35 hrs. from Amoy, and I soon learned from Dr. Morrow that all my dear family were well. I went to Macao to see them on Tuesday evening in the Fenimore Cooper, and spent Wednesday & Thursday in Macao. How pleasant was the meeting, those know who have been long separated; God had answered all my prayers for their health & safety, had provided them a spacious house, and loaded us all with benefits. The inspection of the curiosities brought with me, furnished amusement during the two days I was there, and their distribution gratified the givers & receivers in an equal degree.

I came up today in the Mississippi, and reached Canton at dark, the whole party soaking wet from exposure to a furious squall.

Thus ends my expedition to Japan, for which praise be to God.

Reference

George Staunton, *A Complete View of the Chinese Empire*, published in 1789 by G Cawthorn, British Library, London.

Robert Morrison, *A View of China for Philological Purposes*, published in 1817 by P.P. Thoms, Macao.

Andrew Ljungstedt, *An Historical Sketch of the Portuguese Settlements in China*, first published in 1836 by James Munroe, Boston, reprinted in 1992 by Viking Hong Kong, and also reissued by Elibron Classics, 2006.

W.H. Medhurst, *China: its State and Prospects*, published in 1838 by John Snow, London.

Memoirs of the Life and Labours of Robert Morrison, written by his widow, published in 1839 in 2 vols., by Longmans, London.

S. Wells Williams, *The Middle Kingdom*, in 2 volumes, published in 1848 by Wiley and Putnam, New York.

Bt. J.W. Spallding, *The Japan Expedition, an Account of Three Visits to the Japanese Empire*, New York, 1855.

Commodore Perry and the Opening of Japan, Narrative of the Expedition of an American Squadron to the China Seas and Japan, 1852—1854, compiled by Francis L.

Hawks, first published in 1856, and reprinted by Nonsuch Publishing, U.K., 2005.

Laurence Oliphant, *Narrative of the Earl Elgin's Mission to China and Japan*, in 1858 in 2 vols., by William Blackwood, London.

The Marquis de Moges, *Recollections of Baron Gros's Embassy to China and Japan in 1857-58*, published in 1860 by Richard Griffin and Co., London.

Sherard Osborn, *A Cruise in Japanese Waters*, published in 1859 by William Blackwood and Sons, Edinburgh.

Rutherford Alcock, *The Capital of the Tycoon*, published in 1863 in 2 vols., by Longmans, London.

Robert Fortune, *Yedo and Peking*, published in 1863 by John Murrary, London.

Diplomatic Correspondence of [American Government], Japan, pp. 947-1060. Papers Related to Foreign Affairs. Accompanying the Annusal Message of the Presidentto the First Session of the Thirty-Eighth Congress (first letter No. 22, Mr. Pryn to Mr. Seward, dated Yedo, May 26, 1862, to the last letter No. 50, Mr. Seward to Mr. Pryn, dated Washington, Oct. 3, 1863).

Lin-Le, *Ti-ping Tien-Kwoh: the History of the Ti-ping Revolution*, first published in 1866, by Day & Son, London; reprinted in 2 vols. in 2003, by Foreign Languages Press, Beijing.

Wm. Fred. Mayers, N. B. Dennys and Chas. King, *The Treaty Ports of China and Japan*, published in 1867 by Trübner and Co., London.

Memorials or Protestant Missionaries to the Chinese giving a list of their publications and obituary notices of the deceased with copious indexes, first published in 1867 by American Presbyterian Mission Press, Shanghai; reprinted in 1967, by Ch'eng-wen Publishing Co., Taipei.

Edward Pennell Elmhirst, *Our Life in Japan*, published in 1869, by Chapman and Hall, London.

Letters and Journals of James, Eight Earl of Elgin, edited by Theodore Walrond, published in 1872 by John Murray, London.

Charles Mossman, *New Japan, the Land of the Rising Sun: its Annals during the Past Twenty Year, recording the remarkable progress of the Japanese in Western Civilization*, published in 1873 by John Murray, London.

J.C. Hepburn, *Japanese-English and English-Japanese Dictionary*, abridged by the Author, published in 1873 by A.D.F. Randolph & Co., New York.

Rutherford Alcock, *Art and Art Industries in Japan*, published in 1878 by Virtue and Co., London.

August Henry Mounsey, *The Satsuma Rebellion, An Episode of Modern Japanese History*, published in 1879 by John Murray, London.

John Russell Young, *Around the World with General Grant*, originally published in 1879 by American News Company, New York; abridged edited by Michael Fellman and published in 2002 by the John Hopkins University Press.

Edward James Reed, *Japan: its History, Traditions, and Religions, with the Narrative of a Visit in 1879*, 2 volumes, published in 1880 by John Murray, London.

Isabella L. Bird, *Unbeaten Tracks in Japan, an Account of Travels on Horseback in the Interior including Visits to the Aborigines of Yezo and the Shrines of Nikkô and Isé*, published in 2 volumes in 1881 by G.P. Putnam's Sons, New York.

Frederick Wells Williams, *The Life and Letters of Samuel Wells Williams*, published in 1889 by Putnam's Sons, London; reprinted in 1972 by Scholarly Resources, Delaware, USA, and also included in this series of 大象出版社 .

Félix Régamey, *Japan in Art and Industry, With a Glance at Japanese Manners and Customs*, translated by M. French-Sheldon and Eli Lemon Sheldon, published in 1893 by Putnam's, London.

Stanley Lane-Poole, *The Life of Sir Harry Parkes, Sometime Her Majesty's Minister to China and Japan*, in 2 vols., in 1894 by Macmillan's, London.

Stanley Lane-Poole: *The Life of Sir Harry Parkes*, in

two volumes, published in 1894 by Macmillan, London.

W. A. P. Martin: *A Cycle of Cathy or China, South and North, with Personal Reminiscences*, published in 1896 by Fleming H. Revell Company, New York.

Alexander Wyle, *Chinese Researches*, first published in 1897 in Shanghai; reprinted in 1966, by Ch'eng-wen Publishing Co., Taipei.

Kanichi Asakawa, *The Russo-Japanese Conflict*, with an Introduction by Frederick Wells Williams, first published in 1904 by Houghton, Mifflin and Co., Boston, and reprinted by Hard Press, Miami, in 1997.

E. Bretschneider, M.D., *History of European Botanical Discoveries in China*, first published in 1898 by Sampson Low, Marston and Co., London; reprinted in 2 vols. in 2002 by Edition Synapse, Tokyo.

Alexander Michie, *The Englishman in China during the Victorian Era as Illustrated by the Career of Sir Rutherford Alcock*, published in 1900 in 2 vols., by Blackwood and Sons, London.

A.B. Freeman-Mitford, *The Attache at Peking*, published in 1900 by Macmillan, London.

Yung Wing, *My Life in China and America*, first published in 1909 and reprinted in 2008 by Earnshaw Books, Hong Kong.

Payson Jackson Treat, *The Early Diplomatic Relations*

between the United States and Japan 1853-1865, published in 1917 by John Hopkins Press.

James Murdoch, *A History of Japan*, vol.III, The Tokugawa Epoch 1652-1868, published in 1926 by Kegan Pall, London.

C.A. Montalto de Jesus, *Macau Histórico*, first published in 1926 and reprinted in 1990 by Livros do Oriente, Macau.

Kenneth Scott Latourette, *A History of Christian Missions in China*, first published in 1929 by Society for Promoting Christian Knowledge, London; reprinted in 1975 by Ch'eng-wen Publishing Co., Taipei.

Captain C. R. Boxer, *The Embassy of Captain Gonçalo de Siqoeir de Souza to Japan in 1644-7*, published in 1938, limited edition No. 276, Macau.

Ernest O. Hauser, *Shanghai: City for Sale*, published in 1940 by Modern Book Company, Shanghai.

C.R. Boxer, *Fidalgos in the Far East 1550-1770: Fact and Fancy in the History of Macao*, published in 1948 by Martinus Nijhoff, the Hague.

J. M. Braga, *The Western Pioneers and the Discovery of Macao*, published in 1949 by Imprensa Nacional, Macau.

G.B. Sansom, *The Western World and Japan: a Study in the Interaction of European and Asiatic Cultures,* published in 1950 by Alfred A. Knopf, New York.

W. G. Beasley, *Great Britain and the Opening of Japan 1834-1858*, published in 1951 by Luzac & Co., London.

C.R. Boxer, *The Christian Century in Japan 1549-1650,* published in 1951 by University of California Press.

James Brodrick, *Saint Francis Xavier (1506-1552),* published in 1952 by Burns Oates, London.

Hugh Borton and others, compiled by, *A Selected List of Books and Articles on Japan in English, French and German*, published in 1954 by the Harvard-Yenching Institute.

George Alexander Lensen: *Russia's Japan Expedition of 1852 to 1855* (University of Florida Press, Gainesuille, 1955).

Philip Henderson, *The Life of Laurence Oliphant, Traveller, Diplomat and Mystic*, published in 1956 by Robert Hale, London.

George Lensen, *The Russian Push toward Japan 1852-1855*, published in 1959 by Princeton Univ. Press.

C.R. Boxer, *The Great Ship from Macao: Annals of Macao and the Old Japan Trade, 1555-1640,* published in 1959 by Centro de Estudos Histórico Ultramarinos, Lisboa.

Henry Heusken, *Japan Journal 1855-1861,* translated and edited by Jeannette C. van der Courput and Robert A. Wilson, published in 1964 by Rutgers University Press.

Hsin-pao Chang, *Commissioner Lin and the Opium War,* published in 1964 by Harvard University Press.

Memoirs of Protestant Missionaries to the Chinese, published first in 1867 by American Presbyterian Mission Press, Shanghai, and reprinted in 1967 by Ch'eng-wen Publishing Co., Taipei.

Grant Kohn Goodman, *The Dutch Impact on Japan (1640-1853),* published in 1967 by E.J. Brill, Leiden.

John R. Black, *Young Japan—Yokohama and Yedo 1858-79*, reprinted in 1968 with an Introduction by Grace Fox, in 2 vols by Oxford Univ. Press, Tokyo.

Grace Fox, *Britain and Japan 1858-1883,* published in 1969 by the Clarendon Press, Oxford.

J. S. Gregory, *Great Britain and the Taipings,* published in 1969 by Routledge & Kegal Paul, London.

Raymond Callahan, *The East India Company and Army Reform, 1783-1798,* published in 1972 by Harvard University Press.

G. B. Sansom, *The Western World and Japan, A Study in the Interaction of European and Asiatic Cultures,* published in 1973 by Knoft, New York.

Austin Coats, *A Macao Narrative,* published in 1978 by Oxford University Press, Hong Kong.

Michael Wise, compiled by, *Travellers' Tales of Old Japan*, published in 1985 by Times Books International, Singapore.

Katherine Bruner and others edited, *Entering China's Service, Robert Hart's Journal 1854~1863*, published in 1986 by Harvard University Press.

John Lust, *Western Books on China Published up to 1850,* published in 1989 by the Bamboo Publishing, London.

P. D. Coates, *The China Consuls, British Consular Officers, 1843—1943,* published by Oxford University Press, Hong Kong, 1988.

Hugh Dearn, *Good Deeds & Gunboats*, first published in 1990 by China Books & Periodicals, San Francisco; reprinted in 2003 by Foreign Language Press, Beijing.

Francis Hall, *Japan Through American Eyes: the Journal of Francis Hall, Kanagawa and Yokohama, 1859-1866*, edited and annotated by F.G. Notehelfer, published in 1992 by Princeton Univ. Press.

Michael J. Moser & Yeone Wei-chi Moser, *Foreigners within the Gates, the Legations at Peking*, published in 1993 by Oxford University Press, Hong Kong.

J.F. Moran, *The Japanese and the Jesuits,* published in 1993 by Routledge, London.

Portugal and Japan: XVIth and XVIIth Centuries, Reflections of Encounters, Review of Culture, No. 17, October to December 1993, published by Instituto Cultural de Macau.

Jonathan D. Spence, *God's Chinese Son the Taiping Heavenly Kingdom of Hong Xiuquan,* published in 1996 by W.W. Norton & Co., New York.

Lindsay and May Ride, *An East India Company Cemetery, Protestant Burials in Macao*, Hong Kong University Press, 1996.

Letters from China: The Canton-Boston Correspondence of Robert Bennet Forbes, 1838-1840, compiled and edited by Phyllis Forbes Kerr, published in 1996 by Mystic Seaport Museum, Mystic, Connecticut.

Detlef Haberland, *Engelbert Kaempfer, a Biography*, translated by Peter Hogg and published in 1996 by the British Library, London.

Jacques M. Downs: *The Golden Ghetto, the American Commercial Community at Canton and the Shaping of American China Policy, 1784-1844*, published by Lehigh University Press, Bethlehem, 1997.

Richard Sims, *French Policy towards the Bakufu and Meiji Japan 1854-95* published in 1998 by the Japan Library, Richmond, U.K.

Elijah C. Bridgman: *Glimpses of Canton, the Diary of Elijah C. Bridgman, 1834-1838*, Yale University School

Library, Occational Publications No. 11, New Haven, 1998.

Engelbert Kaempfer, *Kaempfer's Japan: Tokugawa Culture Observed,* edited, translated and annotated by Beatrice M. Bodard-Bailey and published in 1999 by University of Hawaii Press.

George H. Kerr, *Okinawa, the History of an Island People*, revided edition, published by Tuttle, Tokyo, 2000.

Harriett Lowe, *Lights and Shadows of a Macao Life: the Journal of Harriett Low, Travelling Spinster*, in 2 volumes (Part 1, 1829-1832; Part 2, 1832-1834), published in 2002 for Nan. P. Hodges by The History Bank, Woodinville, WA.

British Envoys in Japan, 1859-1972, edited and compiled by Hugh Cortazzi, published in 2004 by the Global Oriental, U.K.

Alain Pichon, edited by, *China Trade and Empire, Jardine, Matheson & Co., and the Origins of British Rule in Hong Kong 1827-1843,* published in 2006 by the British Academy.

James Bradley, *The Imperial Cruise, A Secret History of Empire and War*, published in 2009 by Little, Brown and Co., New York.

Xiaoxin Wu, edited by: *Christianity in China, A Scholar's Guide to Resources in the Libraries and Archives of the United States*, first edited by Archie R. Crouch and others, this second edition, published by M. E. Sharpe,

Armonk, N.Y., 2009.

Julia Lovell, *The Opium War*, published in 2011 by Picador, London.

Appendix

Miscellaneous Unpublished Correspondence of S. Wells Williams, 1853 - 1854.

[Yale University Library Archive: M547, Box 30 and others]

[D.N. Spooner to SWW: 1853/ 04/ 19]

D. N. Spooner

Apr. 1853

My dear Sir,

Commodore Perry has expressed such an earnest desire to have you accompany him to Japan, that I trust you will excuse my expressing the wish that you will do so if it is possible. It seems to me his mission may be productive of very great good, not only to merchants as such, but to the cause of Christianity throughout the world.

Very truly Yours

D. N. Spoon

April 19th

S. Wells Williams Esq.

[SWW's wife Sarah to Catherine Huntington Williams: 1853/ 04/ 20]

Canton, April 20th [n.d. /1853]

My dear Mother,

We were greatly delighted to get your letter of January, tho' short, it gave evidence that you was able to be about, a little, or rather could sit in your chair & be moved about. I am very thankful for this improvement as we were much alarmed when first hearing of your accident, fearing you might be a cripple for life.

The Mississippi came up to Whampoa yesterday, & we had a swarm of officers to call in the course of the day. Commodore Perry is extremely anxious to get Wells to go as Interpreter to Japan, & indeed depended entirely upon his services from hearing in the U.S. that Wells was very anxious to go. I am sure I do not know whom this idea should have come from, for tho['] Wells has often said he should like to go again to Japan he never did it in a way that could be made use of as wishing to be made a member of the Expedition.

I feel quite put out that the Commodore should have waited till the last minute (thinking for granted Wells would jump at the chance) before asking his services. If he had only written him 2 or 3 months since, he might have made some arrangements; but now it seems almost impossible. Tho' I fear Wells will be persuaded to go at last, everything is unsettled yet, but the Com. is very urgent, & everyone here says he must go. The only thing I see in favor of it is it will be a good thing for his health such a health to him such a fine summer cruise escaping the hot weather here as they would not be back till September. John I hear from the Com is engaged as a worker of the telegraph on board the ship, to surprise the Japanese.

The Bridgman arrived 2 or 3 weeks since after a very fine passage by way of California only 165 days.

They are both in excellent health & spirits; it is a great treat for me to see them after so long looking from them. They go to Shanghai again in the first steamer which leaves for the north, & Mrs. B. is full of zeal for her girls school — she got several donators for girls tuition when she was in America & is greatly encouraged that she can go on without calling it all upon the Board for its support.

Your $30 was received by last mail, & Mrs B. sends many thanks for our kind interest. Mrs Martin sends about $50 a year & seems to keep up special interest in China, notwithstanding it was predicted she would soon tire of it. I cannot write Martha this month, for we are in great confusion of visitors calling, & getting a box ready to send Mrs. Murton, so I have only a few minutes at a time to write a line or two.

The New Missionaries arrived at Hong Kong this week. Mr & Mrs. Hastwell they had a long passage 163 days, as long nearly as the Bridgmans were coming via California.

Wally & Kate are writing letters by my side & between them asking for everything at the desk, & being called off in little while, I scarcely know what I am writing about. Olyphant requires two or three to watch him while he is awoke for he creeps like a little spider from one thing to another as fast as he can go, & is constantly getting in misdoing, he is full of fun & frolic & goes crowing about us as if it was the greatest fun in the world to be turned into a moss.

The news has just come from the north that the Rebels have taken Nanking, & another large city near, so the Chinese here begin to fear the present Emperor will soon be to give up to a New Dynasty; they are dreadfully frightened at the North.

With the love to all & Sophy in particular.

I am truly & affectionately yours

Sarah

You got my [one illegible word] letters from Fred this mail.

[W.S. Walker to SWW: 1853/ 04/28; SWWOC[60]]

U.S. Ship, "Saratoga"
Macao Road, China.
April 28th, 1853.

Sir,

I learn from the Commander in chief that you will probably be ready to join this ship by the 6th or 10th of next month. My stay at this port is limited to the 15th, and I am ordered to sail the say after. Will you be pleased to inform me at what time I may expect you at Macao in readiness for departure.

I am, Sir, Very respectfully
Your obdt. Servt.
W.S. Walker
Commander, U.S.S. "Saratoga"

S. Wells Williams, Esq.
Canton

[SWW to his friend James Dana: 1853/ 04/ 30; SWWOC]
Dear James,

An answer to your kind letter of July 11th was begun sometime ago, but something interrupted the completion of it, & now comes a parcel of these little books, & two more notes of July 26 & Nov. 2 by Rev. C. Jones. The box came safely in the "Siam", and the dogs & wagon have been hauled along over the house, verandah, hong & garden, until the latter has got well worn. The team has been left from distraction by withholding it except when visitors come. We are much obliged, for our kind remembrance of the little folks, and Walthorth remembers [No] Dana as connected with the white dogs, and thinks he will see him when he goes in the large ship to Whampoa & England & America and other small towns in their neighborhood. The books are quite beyond him, he having hardly learned his letters yet. Kate is eager, restless, impatient & winsome, and if Gos directs her heart & blesses her training, she will make a useful person. What anxious blessings these darlings are, especially in a heathen land, or in any land, how much faith, prayer & patience are required to guide & govern them. May we both find our children loving God & his service.

Mr. Jones gave us some additional notices of your family & townspeople. I was glad to hear that you are getting on so well with your zoology & crustacean. In the large collection of fishes I sent you some years ago from Macao, which I believe, got lost somewhere, there were many of the crabs, [crangons], shrimps, [gammari] & other sorts of this great family of crustacean. These might have assisted somewhat. I have not heard whether there are any persons in the Japan Exn. to collect such things as may occur or not. I shall have little time for such pursuit, I expect, but they will repay as much as anything connected with the expedition.

I have not the clearest idea of their objects of the expedition to Japan, and hardly any one has, I infer from what I can hear from officers connected with it. Something is to be done, and rather unwillingly have I been induced to join it. Why you all should have supposed that I was to go with this fleet, I cannot imagine, & it has been a difficult matter for me to leave my printing office & works in hand to go. I am willing to help in all good undertakings, but the inception & object of this are doubtful. I am engaged for 5 [months] only, by which time I trust the path of duty may be made clearer, & the probable hap of the whole enterprise more evident than at present.

Numbers of the Am. [Inb.] [Hyive] come, & I look over the pages with interest; your publication of crustacean gives you matter for it, I see. I am often impressed at the entirely opposite lines of study we have been led into, & that you may see what I am doing I inclose you a sheet of my Vocabulary of the Canton Dialect, now progressing to its middle point. What a contrast to our canal boat exercises, to [Ariskany] swamp after [Habenarios], [Diabarda] & [Cypripedia]! Yet I take increasing interest in such dry studies as these from the hope that this is & other books will assist others to learn the language easily & help them preach quicker. My printing office is under the same roof, & I often go from the dining-table to the press to see if all is going on right. My copy does not lie very long on hand before it is in type, which is a disadvantage in some respects, as I have not so much chance for correction. The demands for copy by printers whom I am paying myself & cannot afford to have idle, [also] not quite so urgent as to get out of the way of a locomotion you hear thundering behind you, but they are longer continued and are not so easily avoided. However, I keep good health, & this cruise to Japan may prolong it. We have only to keep on during this sojourn, regretting nothing but sin, & trusting for joy & peace behind this scene of probation.

Make out best regards to Mrs. Dana, also to Prof. Sillieman & his son, & their families, and Macy if he is in N. Haven. Tell Awing, I haved paid his mother $25 as he wished.

Ever & affly.
Your old cham
S. Wells Williams

[Dr. B. J. Betteleheim to Commodore M.C. Perry: 1853/ 05/ 29; Box 30]

Napa, May 29th, 1853.

Commodore M.C. Perry

U.S. Navy.

Sir,

I deeply regret to have to inform you that the Chinaman sent into my house is not ill from sea sickness. He is a confirmed opium smoker. And as it cannot but be highly injurious, if the fact — which sooner or later it must — be divulged among the Lewchewans, I am in the necessity of begging you to let him forthwith be taken on board his ship. He told me his opium was there.

In consequence of this unfortunate, or, perhaps I should rather say, fortunate discovery — not to mention the extremely bad & almost unintelligible pronunciation of the man, in my opinion, he is totally unfit for his position of spokesman, copyist, or any share whatever in your great Commission. And if you should not deem it of sufficient urgency to despatch a ship expressly for the purpose of bringing from China two excellent speakers of the mandarin dialect I cannot hid it from you that is in my humble opinion the success of your Expedition is so intimately connected with perfection in the Interpreter's department, that I feel it my boundless duty humbly to suggest that the first ship proceeding hence to China be charged with the Commission of procuring two accomplished speakers of the mandarin dialect to arrive before the Squadron's arrival at Japan. The men would be best chosen from among those brought up at Macao, who speak latin. This would be an additional important advantage, as it is but likely the Japanese government had Dutch spies, who understand the English & Chinese, but are very poorly off for latin.

I have no doubt but the Bishop of Victoria, Dr. Bridgman, Medhurst & similar men would willingly part with their own, or otherwise find ways speedily to accommodate us with suitable men, and I am equally persuaded Dr. Williams' views will coincide with mine in the aforegoing suggestions.

I have the honour to remain

Sir

With profoundest respectfulness

Yours most humble & obedient servant

B. J. Betteleheim

P.S.

In case the man cannot be removed to ship, I must beg a person should be sent to wait on him, as he at times appears delirious & is scarcely able to walk without being supported. Some coffee also would do him good; we have none. If he be removed—which is best—some warm covering should be sent him, to be wrapped round him while in the boat.

[Dr. B.J. Betteleheim to Commodore M.C. Perry: 1853/ 05/ 31; Box 30]

Napa, May 31st, 1853.

Commodore M.C. Perry

U.S. Navy

Sir,

I have the honour to inform you,

That the Squadron has now an officer on shore, in a position eligible from a combination of many advantages.

That I have been asked by the Loochwan authorities (of course in strict privacy) whether the doors of the palace should be opened to you in a friendly way, and, of course, I answered in the affirmative, assuring them—as I sincerely could—that yielding on their part will prove most conductive to their real advantage.

Notwithing they will try—so their interpreters publicly informed us—to wait upon you to—day concerning the palace meeting & other matters. And it is likely the chief manadrin, already introduced at yours [an illegible word], will himself be the visitor. He could scarcely be [an illegible word] admission, as he might think himself thereby [an illegible word] to a similar stop when we proceed to Shuy.

In my humble opinion

If Dr. Williams [Samuel Wells Williams] would manage to be at the office, while the Loochwan Authorities approach the Susquehanna, in his absence the visit might be declined, as there was none to interpret. This palliative however has a drawback, & that is, that the native Interpreter himself speaks English tolerably well, of which he gave yesterday astonishing evidence before your excellency.

Might it not [an illegible word] to have a board or two hung down the sides of the Susquehanna as a scaffolding, & set a man or two to put on a little paint, or drive he nails deep in &c. so that the ships might outside appear "under repairs", & unable to admit a high personage on a diplomatic visit.

In a note given me by one of your officers I read: Ask Dr. Betteleheim ["] how can be compensated for his services, which we want for the interest of the Squadron?" My dear wife likewise told me, she had been asked whether she would permit me to accompany the Squadron to Japan? Both questions naturally presuppose your adopting my humble self as member of your glorious expedition. Still [as] in the event of my going to Japan I should be obliged to give definite information [as] to my connexion both in China & England by your mail of after to-morrow, besides making several other arrangements, I feel necessitated hereby to beg in my plain simple way, your word or wishing on this my destination, if such it is to be.

In this case I should also beg permission to have brought over by your return mail an American young man of good report, either as servant, or housekeeper. Indeed I should beg this in any case, as my position here would be greatly strengthened by engaging American aids. I am ready to salary him according to his station in society. And I crave your cooperation at your agents in China in order that such person might be speedily found. I shall certainly not fail to request my friends to be on the look out.

May I beg you to please to order every moving a boat to take me to the office, when Dr. Williams desires me to be, & to bring me back in the evening, till the horses will come forth.

I have the honour to remain

With profoundest respectufulness

Sir

Your most humble & obedient servant

B. J. Betteleheim.

[Lewchew Regent's petition to Commodore Perry: 1853/ 06/ 03; a formal Chinese letter; pp.2]

[The English translation of the above letter, written in SWW's handwriting]

A prepared petition. Sháng Ta-mu, high minister superintending the affairs of the kingdom of Lewchew at the prefecture of Chung-Shan, hereby (acknowledges the) receipt of orders, & urgently begs to communicate on an affair.

On the 28th ult. I received Your Excellency's injunction, "that on the 6th inst. June, you wished to enter the royal palace at 10 o'clock a.m. to return thanks", &c.; or that "since the Queen Dowager's illness had increased in violence and her disease had not been checked by medical treatment, you begged to be allowed to go to the Heir Apparent's Palace to tender thanks." Through Bettleheim who was on board your ship yesterday, I have to acknowledge Your Excellency's orders, to the effect "that I have not seen the Crown Prince, but it will not be inconvenient to go into his palace; and further the English general last year went into the royal palace, which I also wish to do."

On receiving this, I immediately memorialized respecting it, & have now received the commands of the Ruler of the Kingdom to this effect: "Ever since the English officer entered the palace, the Queen Dowager has been ill, & her disease has not yielded to medical treatment, so that it is much to be feared that if His Excellency enters the palace her illness will be increased, & may be difficult to cure. The Regent must therefore arrange the matter properly and entertain him at the time; but it is important for him not to enter the palace. Respect this."

I have reflected that I have enjoyed for my whole life the favor and equity of my country as if I had not been a relative (of the king?). To arrange this matter, therefore, in such a way as not to cause any grief to the sovereign, is my peculiar duty as an officer. Moreover, when the English went to the palace last year, the Queen Dowager was not ill, while now her sickness is not yet cured; so that the circumstances are dissimilar in the two cases, and how much more so when I have received notice that personally she is apprehensive of running a fatal risk. Wherefore I again intrude on Your Excellency and humbly beg that you will confer an additional kindness on me, & come to my official residence there to have the ceremony (of an interview). If I do not obtain this favor, I cannot again behold my sovereign receive his orders. My life and death are in Your Excellency's hands alone, as you assent to this or not; & I beg that you will, on the one hand consider the true circumstances of the Queen Dowager's illness, and on the other compassionate my dangerous position personally, giving them your careful thoughtfulness. If you consent to that which I request, then will all parties join in wishing you unabated happiness. An urgent petition.

Hienfung, 3d year, 4th month, 28th day (June 4th, 1853)

[SWW to his wife Sarah: 1854/ 01/ 13; SWWOC]

Jan. 13. 1854

My dear Wife,

I send you in the basket, the two plates and some pamphlets I found here for me, recd. per Layington; also my penknife, which will be useful for mending Mally's leadpencils, if for nothing else; it has become so dull that I have bought another one.

The mince-pies and crullers were recd. with the thanks of all who got them, and Mr. Mounsy, Mr. Bent, Dr. Smith & the Comre. made inquiries after you. John [SWW's younger brother] is much better of his cough, & says it does not trouble him at all, tho' it has not gone entirely; he looks pretty well, and is doubtless much better than when he was in Canton. It is said that the Commodore intends to return by way of Shanghai, so that it may be well for you to send a line there, by some opportunity two or 2½ mos. Hence, and also John's letters, if there be some one going up who will take them, or it is hardly worth while to put them in the Post Office.

I send an order for $50 on Mr. Pustan, which Mr. Bonney can get when he wants it. This and, the money in hand will be enough for expenses up to Feb. 25 or so, and I suspect by that time other sums will come in, so that you will have enough.

Mr. Morton & family have got a passage to Lewchew in the ~~ a ship going to California, which leaves him there for $800 or $1000. I have not hitherto seen him. There is little news here. Mrs. Hunter is still here, & the operation on Jenny at present bids fair to prove successful, so that Jennie will be able to walk easily. I have not yet seen Hunter, & may not be able to do so. We are to sail to-morrow morning without doubt, is the present report, mail or not; & I must sleep on board to-night, having remained two nights at Mr. Johnson's.

Mr. Bent has taken his map such as it is, & paid me three dollars; I have paid the man two, & when he brings your picture frame, please pay him another one, which will settle that account.

I have just returned from a visit to all the steamers with Mrs. Johnson, & hear that the squadron goes to sea early tomorrow morning, so that you will not have any chance to hear from me again. There is to be a storeship sent over to Shanghai, by which the officers hope to get letters, if the mail is not in before they leave, Perry having left orders for his shippers or agent to send his letters up there. John has no note to send except this one to Mr. Whipple; his love and kind words are all included in mine.

I have seen Mrs. Purtan, who hopes to go up next week, and spend a short time at Canton; she wishes you still to make her a visit here. The chances are that you may be able to accept it. I have seen Mrs. Smith & the Bishop. Mr. Morton leaves on Monday next direct for Lewchew, so that you may still be able to drop me a line by him, & the chances are that we shall not leave Napa before he reaches it.

Goodbye, dearest, & keep up good heart; rejoice the

Lord all day long, & trust in him when you [one word missing on account of paper torn] pray with Wally, & teach him too, not to pray to you but to God himself. You will be glad years hence, perhaps, that you have done so, now. Try, I beg you.

Dr. Gambrill is here at Mr. Johnson's very sick with the dysentery, & his symptons today are not so favourable. He belongs to the Plymouth.

Kiss the childen from papa. Love to Mr. Bonney, Dr. & Mrs. Parker, French, Happer, Vooman, Mrs. Brewster, & kind regards to, the Nye's, Purdon's, 'c.

In love ever Afftly,
your Wells

[SWW to his wife Sarah: 1854/ 01/ 31 & 1854/ 02/ 1; SWWOC]

Napa, Jan. 31, 1854

Dear Sarah,

We have been here now well on to a fortnight, and will be off in three or four days to Yedo whither all the sailing ships started this morning with a fair wind. We left Hongkong on the 14th and got here the next week Saturday evening, without any seasickness on my part, and not much unpleasant weather. The Lewchewans send off a deputation to salute the Commodore next day, & except three days of windy weather in which it was impossible to go ashore, I have been to see the authorities almost every day, scolding, exporting, advising & coaxing, them as seemed most likely to obtain our objects. The common people have become far more social, and gather about us when we stop, and the children exhibit no fear of us as they did. We have compelled the rules, in this way, to receive pay for the house, and made them give receipts for the money—great achievements, you will say, yet they were attended with some difficulty & the exercise of patience. However, a good deal of progress has been made both with rulers & people, and the dozen old weatherworn sailors who are to be left here in a sort of half pay half sick way to look out for the [walshed], will serve still further to familiarize them with us and our ways. Some of the purveyors have learned a good deal of English from the men left ashore from time to time, and as we go along the streets, we now & then hear little boys chirrupting out, "How do' you' go?" "Good morning" "American," & other phrases, and then hopping back into the oblivion & dirt they emerged from. Some of the harridans in the market places seem disposed to sell an articles occasionally, but they are afraid of such public contact with the foreigners, and rather avoid it. Yet a good deal of progress of the kind we desire has been made, and I hope it will meet with no hindrance.

Dr. Bettleheim sends his family over to Shanghai by the Supply, & two Chinese he has with him, himself remaining till Mr. Morton & family arrives; perhaps the Robinia may come in yet, for Capt. Sinclair is willing to take all their household. The Chinese came over here last summer, & have done him no service, but have drank large potations of saki, while Dr. B. himself shows them the example of taking small draughts of spirits. I really can't yet decide from all I hear how Bettleheim is regarded by the people here, but they contrive to leave him in solitude most of the time, except a servant or two to cook & run errands. I believe no native women has ever entered the houses.

The climate since we came has been mild, & even warm, so that woolens are uncomfortable; the natives have a jacket or two more for garments, but not much difference is to be seen. No fires are even used in chimneys, & I suspect most of their fire is in the cookroom.

Dr. Morrow, John [SWW's younger brother] & Heine, & their mess, have gone ashore to live; the first & last have gone on a trip northward to see about coal reported by Lieut. Whiting. John has about recovered from his cough, & it gives him no trouble. He is idle as he well can be, & has been set to collecting Lewchewan words as a pastime, in which his progress is slow, as you may well believe, seeing his turn is not decidedly for making dictionaries.

I find I have forgotten some things, among which my book to put plants in, a lot of Peking Gazettes, and some more History of the U.S., all of which I had intended to bring. There are not many things a man has which will not perhaps be wanted on board ship sometimes or other, so he cannot always tell what will be needed first.

Feb. 1. [1854]

To-day four of us were dispatched to Shui to hand in a letter to the Regent, & in order to make it impressive on the people and rules at the capital, about a hundred marines & musicians were sent with us. The hour was early, & the visit entirely unexpected on their part, and it was therefore a great surprise & treat to the natives, to see the show and hear the music; but to the authorities nothing could have been more startling. They supposed, I had nearly concluded on seeing them, that our intentions were nothing less than to take them prisoners on board ship; they hurried out of the house in the greatest trepidation, while the evolutions of the marines were surprising and threatening, so that in fact I had some trouble in quieting their fears. The regular interpreters were not at hand either, so that our communications were slower than usual. The Regent acted as if out of his wits, got up & sat down, took us by the hand to detain us, and did what he could to prevent the Commodores visit on Friday.

After having been here so many months as we have, it is strange that these islanders have so much fear of us, and will not believe what we say, even after the reiterated statements of our friendly designs. One Lewchewan has however trusted to us, & has come off to the ship by himself, desirous to go off with us. He swam off to the Vandalia a month ago, Capt. Pope seemed to think it was time he fared better, & is likely now to see something of the world. He talks intelligently about his own island & its customs, and is no simpleton, so perhaps we may yet make something of him. We have one Japanese aboard & a Lewchewan will add to the variety, & give John something to do in collecting words. The day was very breezy and clear, & every one who went to Shui was charmed with the jaunt; but our hosts were doubtless less pleased.

The Supply has at last cleared herself and will soon have in her ballast, and be off. I am also erelong to move all my traps and followers into the Powhatan, where the Commodore will move when the squadron reaches Yedo. The boat we came here in is to be put at the service of Mr. McLane and I suppose he will like to take a few excursions in so desirable a yacht; this compels the Commodore to move out into another ship, greatly to his own inconvenience & discomfort of others. I am likely to be quite as well, & a little better off perhaps; than when here, for some of the officers here are not the most pleasant, & perhaps some there will not prove more so. There are many intelligent & sociable men in the squadron, but no mess that I have seen equals (i.e. suits me like) the Mississippi's.

I have been away from you three weeks, and as you would like to hear how I get along, so do I greatly wish to know how all the loved & loving ones in No.2 Mingqua's hong are. I trust you are well, & daily commend you to

God’s goodness & care; and even if what we call misfortune comes on you or me, let us still so see God’s hand in it that we shall regard it as good. Mr. Maury requested special remembrance to you, this morning on shore, where I saw him a few minutes. Also Mr. Upshur & Dr. Wilson, and Upshur sends a kiss to Kate, whose sharp eyes have taken his fancy, and her name pleases him too.

I am uncertain when this will reach you, and know still less when you will have a chance to send anything to me. If the Robinias comes in after we are gone, that letter is likely to be a long time seasoning before it is read, inasmuch as we shall perhaps be detained at Yedo some time in any case, & longer if we succeed, than not. The Commodore has not yet found time to put up his printing press, and I don’t know when it will issue his bulletins. Perhaps in Yedo bay.

John contents himself with sending his love and good wishes, too lazy to bestir to even answer Robert’s letters, and tell him what luck befalls him in this distant land, when the friends at home would be so happy to hear everything. I find my time very fully occupied in reading, study and all manner of errands on shore while we are here, so that I find it travels on to eternity pretty fast. I have written 70 pages letter sheet of a translation, which will make a torelably readable story, I think, when trimmed up. The teacher enjoys the sights he sees here very much, making his notes on people & things. The great crowds of children all of the same size seen in the streets has[sic] attracted his notice as it has everybody else’s, & now that they swarm around us, their size & countenances are more intelligible. The improvement in intercourse is very great, and by degrees we shall perhaps get into the houses of the better sort, and see somewhat of the [town]. As I hear that we shall probably stop here coming back, I have bought nothing yet, for my accommodations are not very spacious. If I do not come back to Napa, there will be great loss.

I inclose a letter for Fred partly written, preferring to send this than none, as I have now no time to finish it in full. You can fill up with news from China, and perhaps I shall be ableto send another sheet from Yedo, when the Commodore sends the Susquehanna down to Hongkong, and tell him something more. You can read the notes for Nye & Parker, & then seal & send them in, as they are additional to these few lines.

You & the babies are with me every morning & night, thought of, prayed for, & longed for. May God watch over you all, & bring us together again in his good time, to praise him for his mercies. Please remember me to the Vroomans & Mr. Brewster, to Happers, Trench, & all their families. I suppose Dr. Ball has left, or will before you get this. Kind regards to Mr. Hitchcoch, and Mr. Dent & other friends. The Commodore & Adams stay aboard the Susq[uehann]a. till they reach Yedo, and then they come to the Powhatan, where I am now housed just as I was in her in a little cabin on the taffrail. John is all in a dump from the way he is turned over; I suspect he wishes himself back to U.S. a dozen times a day. Dr. Morrow sends his regards.

Good bye, dear loving wife, & may God bless you with his Holy Presence & counsil, & make us a means of good to others.

Ever your's

Wells.

Steam Frigate, Powhatan

Feb. 5. 1854

[Shang Hiung-hiun to Perry: Hienfung, 4th year, 1st moon, 5th day; Petition; SWWOC]

A prepared petition.

Shang Hiung-hiun, Superintendent of Affairs in the Middle-Hill prefecture in the kingdom of Lewchew, high minister, & Ma Liang-tsai, Treasure, &c.,

Hereby petition in reply. We yesterday received Your Excellency's official communication, and ther[e]in find it stated;

1st ("You say" that) the officers of Lewchew treat the foreigners in a very unfriendly manner, and do not regard truth in their intercourse with them:" Now we have carefully looked into this, and we are sure that from the first we have strictly enjoined on our subordinates and followers that whenever they meet or see your officers or men, they observe the greatest respect and regard, and never exhibit ill manners in any way. On receiving your Excellency's commands, we were alarmed and surprised in no small degree; but we will hereafter issue strict orders to pay attention to this thing without fail.

2nd. (It is observed "that) your officers cause the people to [keep] from us, and to regard us as their enemies, so that thereby dealings between them and us are prevented; and further, that in supplying provisions you have not furnished what was asked for": We, the Regent, &c., have long since strictly enjoined it on the lower classes and people, that whenever they meet the Americans they must not violate propriety & run away; but the low people in the market fear the power and dignity of the visitors and run away of themselves, but we will now enjoin it on them not to act in this manner any more. Furthermore, the traders in the market are all of them women, & the articles they sell are trifling, such as rice, grain and miscellaneous things, and they never use any silver coin in their transactions; the Americans would therefore be unable to supply themselves with provisions in the market, and official purveyors have been appointed for this reason to attend to this matter. Moreover, this country is a little out of the way island in a corner of the sea, poor and unfruitful, producing nothing of consequence; and since the arrival of your ships in May last year, one after the other anchoring in these seas, the things required by them have been very numerous indeed. At present, the policemen and petty officers have sent persons all over the country to seek for things wanted in the orders, and though they may be very difficult to execute, there is no help for it, but if the tale of articles is reduced, orders shall be issued to these officers to exert themselves to supply them all as they are wanted.

3rd. ("You say" that) you do not, in what you agree to and promise, regard the truth or heed your words, but practice much falsehood, and which shows your impertinent disrespect." When we had the consultations the other day which your Excellency directed to be held, we had every desire to be "tremblingly obedient"; but the matter in hand was a difficult one to agree to, if we trust that you will not in this impute disrespect to us, for how can we cherish any impertinence or disrespect? We humbly think upon your Excellency's vast kindness and commiseration for the poor, and consideration for those who have erred, always exhibiting humanity & kindness, to which goodness we look up.

As your Excellency has now stated these several particulars in your communication, we have respectfully replied to them in order, begging that kindness may be manifested towards this little country, and that you will condescend to regard it with compassion, and we shall be deeply moved at such incomparable goodness and regard.

A prepared petition.

Hienfung, 4th year, 1st moon, 5th day.

[SWW to Dr. Peter Parker: 1854/ 02/ 04/1854; SWWOC]

Napa, Feb. 4th, 1854

Dear Doctor,

After a fortnight's stay here we leave for Yedo in a day or two, to see what turn our Expedition will take, and what use God is making of our agony in opening up this Oriental world. To me the result is as evenly balanced, that our next or two seems like entering a Newfoundland fog, where the chances of running on a berg or over a smack are equal to steering clear, and running thro' safe.

Here we have found a great improvement since last August, the people have became more friendly and familiar, & do not shun our presence; the police do not dog our steps and drive us from the people who may be inclined to prolong a visit, and children have begun to speak a few English words, & the purveyors to hold communications. Thus some progress has been made, & hitherto no evil conduct has given the natives no cause of complaint. One native has trusted himself to our kindness, having first swam off to the Vandalia, & been refused admittance, & then came to the flag-ship in a canoe about a week ago, where he is now, hidden among the machinery. He wishes to go to Yedo, U.S., China, anywhere where we sail to, & the Commodore means to take him. It is strange to see how a change of dress transform a man & our associations of him, as instanced in this Lewchewan, who appears like another race in his foreign costume. He seems to be of a fearless nature, & the information derived from him thro' Bettleheim hangs together so well as to show a shrewd mind. I hope something may be made of him productive of good to the island.

Mrs. B. [wife of Dr. B.J. Bettleheim] & her children go over to Shanghai in the "Supply", where she will remain until Bp. Smith gives some advice respecting her movements. The Dr. can get off in a vessel better, if she is not here, for no man-of-war can easily take a family. His assistance will be needed to start Mr. Morton in the language and other things, and the house is not large enough to hold two families. In missionary work there is no improvement to notice, and the people keep away from Dr. B.'s house entirely, & refuse to hear him long in the streets. It is strange to see a whole people so controlled and restrained from the natural emotions, & impulsies of the heart as this is, afraid to ask for medicine or relief, or her a new doctrine. The scarcity of sick & diseased people, & almost entire absence of beggars, is very noticeable after having lived in China.

The Commodore desires to be kindly remembered to you & Mrs. P. He has been quite well since reaching Napa, & has been at work as usual. No one can be idle that has the direction of such a squadron as this, I am to go up to Yedo in the Powhatan, which he joins as soon as he reaches Uraga; this change is an agreeable one to me in prospect, & I hope will be in function as pleasant.

I am now quite well. My clerk, as he is called here, enjoys himself much, & has written some verses respecting Lewchew. His most compendious observations is, "that the Lewchewans are mannerly & kind, the women ugly, & the children all of size." He is looked up to by the literary people here in wonder, & treats them kindly in return.

With affectionate regards to Mrs. Parker, & to yourself.

As every truly Your's

S.W. Williams

[SWW to Gideon Nye[61]: 1854/ 02/ 06; SWWOC]
(Found Dec.2, 1881)
My dear Friend,

We have been here now over a fortnight, and during that time have been mostly engaged in taking in coal & refitting the ships, four of which have already gone ahead, leaving the Commodore to follow in the steamers. We have had considerable intercourse with the rulers & more with the people, who are rapidly becoming familiar with us & our manners & money, so that there is a prospect of obtaining all necessary supplies of provisions, & probably other things too. Some orders for lacquered ware have been executed here, and silver ware of an inferior workmanship has been done to order on a considerable scale. The familiarity of the children in the streets is one of the most apparent signs of progress, for, while they ran away from us before in every direction, they now come about us and seem under no restraint. The girls are mostly kept out of the way, but there are women enough in the markets and highways to satisfy the strangers that no country shall show uglier faces of the females than this islet. They are grimed in with sorrow, labor, & exposure, until the skin becomes like leather, wrinkled & soiled till it is useless.

The Commodore went to Shuri last Friday with a large escort of a hundred marines & musicians, and 20 officers, where he was entertained by the Regent, and feasted with a dinner of 15 or more courses, the cuisine of which was quite palatable; some of the dishes, were novel, & none of them disagreeable. The Regent received us in the palace, but he would not bring out the Prince to see us, nor gratify us with a sight of any other part of the building than the room we were ushered into. This room is a dark apartment, covered with thick carpeting of mats, and ceiled with red varnished wood, as were the sides of the apartment half way up, the rest being slight drawings on a dull paper, making the whole rather gloomy. The building altogether looks deserted & waste, but kept in tolerable repair, as well as unpainted wood can be, for no one has seen any painted wood, all is varnished.

The country is charmingly picturesque & beautiful, not only in the alternation of valley & hill naturally, but also in the high state of cultivation it is mostly brought under, making it look like a garden. It is not much terraced though fields are seen on the hillsides, but it is generally covered with crops, ripening, reaping, growing, or planted, according to the season.

The American flag floats on shore near the coal-depôt, which has been purchased, and is to be left full in charge of a party of weatherworn old seamen, who reside in a house hired by the year, situated not far off. They will be comfortably housed while we are away to Yedo, and on our return will be relieved by others if they do not choose to stay. The amount of money left here by the squadron since last May is exercising its influence, & the sailors have begun to learn the way to trade for their own supplies; so that what with buying & renting houses, exploring parties sent here & there, and dealings in the bazaars, we are in a fair way to make an impression on this secluded people, a good deal yet remains to be done, however, which may not be done as well as it has been begun. We start for Yedo very soon, today perhaps.

With best regards to Mrs. Nye, Mr. & Mrs. Munroe & Mr. & Mrs. Purdors, all of whom I hope are well, (& have as pleasant weather as we have had, too).

As ever truly Yours

S.W. Williams

[SWW to his wife Sarah: 1854/ 03/ 11, 03/ 16, 03/ 19; SWWOC]

Bay of Yedo, Off Kanagawa
Powhatan, March 11, 1854

My dear Wife,

Who is ever in my heart, as I am her's, must have the first salutation from this land of Japan, that I have ever sent away. I have before written letters here, but taken them myself; & now I send you all the joy & peace which the tidings of good health, quiet temper, abundant occupation, & on the whole agreeable position, can give you. I have been absent from you two months, and since I left Lewchew in this ship, five weeks ago, have really enjoyed existence, feeling well all over. I can trust in God that you are also kept in health & quiet, and the darling gives around you give you pleasure, occupation, care, and mother's joy, in their health & amusements. Yet, even if I knew now that you had suffered; or they had ailed, causing you double sorrow & pain in their sorrow & pain, I would ask God to enable me to rejoice & feel that he had done us good. I can humbly & joyfully leave you all in his hands; & do so every day, our prayer's meeting in her ears.

Com. Perry had his first interview with the Commisoner, whose name is Lin in Chinese, & Hayashi in Japanese, on Wednesday last, the 8th of March, at a place called Yokohama, where four or five houses have been erected for the purpose, and where the presents are to be displayed. The interview had been delayed nearly a fortnight by the endeavor of the Japanese to get Perry to return with his ships to Uraga, where they had made preparations to receive him & doubtless expected to bring him. He refused to go down the Bay, after sending the

Vandalia down there, went up within 15 miles of Yedo opposite this town of Kanagawa, where we are now. The Japanese now agreed to have the meeting here, & brought the houses all the way from Uraga, 30 miles, & reconstructed them on the beach opposite our fleet. We are much safer here, and not liable to delays from high winds preventing our going ashore. The weather has been charmingly cool, and I am greatly impressed by its freshness & clearness; and am pretty sure that I have recovered my precious strength.

March 16th .

There was to have been a meeting between Perry & the Japanese commissioners to discuss some points in the treaty which they have prepared, but which the Commodore objects to, since their object is to confirm us just as the Dutch are at Nagasaki, and this we had better resist at once. The people near our house on shore are as friendly as they can be, and those living in this vicinity are equally kind and civil. Dr. Morrow & I took a ramble a day or two since about 4 miles over the hills & valleys to collect flowers, & we were everywhere trusted politely. The weather has been boisterous & rainy, so that no one has left the ship who could stay on board. The presents for the Japanese government are all on shore, & the railroad & telegraph are now being put up at the house where the interview was held; the lot of agricultural implements is operating out, and may perhaps suggest something new & useful to them, but I believe doubts, since most of them are labor saving, and labor seems to be abundant. The rail road, engine & cars are all on a miniature scale; the car has about 15 pair of seats, all covered with damask, the floor excepted, the windows slide up & down, and all that is wanted is a lot of dolls sitting on the seats; to render it complete. The wheels are only 9 inches apart underneath, so that you may guess the tiny seats inside, each 5½ ins. long, look cunningly. Altogether it pleases our side as much as the Japanese, and when we get steam up, & a dozen of them straddle of the car, (for no one can get in of course) I think we shall [see] some fun. The telegraph is too mysterious to attract their attention as much as the railroad, the corn-cracker & rice-huller even, are more popular; but Colt's revolvers carry all eyes, & everybody wants one; some 13 have been given away.

On the whole we are succeeding pretty well, though the dough is not baked yet, and may yet be burnt by some bad luck. All these dispositions of things at present indicate success in the main points; and I thank God who orders all things that it is so.

Sunday, March 19th.

How happy to have such an interruption in my letter, which I laid by a while ago to wait for more news to tell you, in getting your welcome letter of China new years. It came today in the Supply from Shanghai, and tell me so much that I have wished to hear; and more than which I cannot now expect to hear till I see you again. I am bound always to rejoice, and your letter makes food for gratitude. To know that you & the children are well is enough for a long time; though much more would give me greater joy. In truth it seems so odd to get letters & answer them from Japan, that I can hardly forbear referring to it. I hear that many are told from Shanghai that the consul at Hongkong has many letters for them by him, which he knows not what to do with. I have not got your letter by [W.] Morton, as he had not reached Lewchew when we left for this port, but I suppose we shall call in there by & by/ I gave your letter to John to read, & have at last induced him to write somewhere. Martha & Dwight were allowed to witness a remarkable preservation—one which seemed to say that God took care of casualties not to be guarded against by human foresight in preserving those exposed to them; while those owning to human carelessness, as at Nowark, resulted in so many deaths.

My "Chinese clerk," as Lo [羅 森] is usually called, inquires for news, & I wish he had a line from his family. I tell him all that you give me which would interest him, & he seems to think that a wife may be made of some use after all, as I show him your letter. I often talk with him about people & things in Canton, & both he and Aman have proved an excellent selection. He is making friends among the Japanese by writing poetry for them, & they reply, as there are many who can read Chinese easily, tho' no one who talks it. Lo has secured everybody's respect by his unform quiet way, and his cleanly manners and appearance; & the other fares well among the stewards and cooks & men, who are not usually very well disposed towards Chinese, of whom we have only 6 on board. I have plenty to do of my own & the Commodore's, & so my time passes pleasantly.

I am sorry you feel lonely, but I can't relieve it yet. Why does not [Huter] move his family to Canton for the winter, and then you could have company near you. I leave all my inclosures open for you to read, & you can seal & send them; this gives you all the items, & saves me some writing, of whc. I have enough, with repetition.

I have been ashore every day for ten days past; and the people collected there have given me so much practice in talking, that I have become quite expert in the language to what I was when we reached here six weeks ago. Everybody comes as if they expected no difficulty in making themselves understood, and therefore I have them all talking in the best idioms, I suppose; & my chief effort is to imitate all the expressions I hear.

From Mr. Bittinger, if he reaches Canton, you will learn much of the general progress of things, & answers to many questions whh. will arise in your mind. I am sorry that he is as generally disliked in the whole squadron. Dr. Smith & Mr. Randolph go soon in the Saratoga, which will be on her way home in a week, perhaps sooner. The officers & crews of that ship are pretty much worn out and sickly; Mr. Wayne may not live to reach home. Mr. Harris is a good deal older looking, and the whole ship's company show the effects of a long confinement in small quarters. It will be 6 or 8 mol. yet before she reaches home.

My Chinese teacher & Aman serve many people in one or another way; & inclose their letter for friends; it has been a great affair for Aman sending off "a packet of dispatches", of whh. he can hardly read a whole sentence in the book style, & I should like to see how he manages in the epistology.

My wardrobe is likely to serve me through, tho' I had not be-lining enough, and a pair of boots would have been a good friend, as the shoes I got of Brewster's are not high enough to me over the paddy fields & roads in Morrow & I have taken many walks in this vicinity, picking up whatever is of interest; flowers, shells, &c.

The country hereabout is exceedingly fertile, & the crops of wheat & barley show the labor bestowed on them, which doubtless gets into reward. However, the people are miserably poor in domestic animals. I have nothing better to send you as the products of the land than a fanny piece of candy, & a few thin cakes, the latter of which are broken by my unluckily slipping when I had them in my pocket. They are however just as good for a taste. Next time, I shall probably be able to produce something more durable for you; but hitherto we have not been able to buy anything, so fearful are the people of doing anything to implicate themselves with us. If we had a good lot of Japanese coin, perhaps more things could be procured, as then the sellers would have nothing to betray them.

The best kind of affectionate remembrances are sent to you & the children by Manry, who wishes each of the young ones to be kissed for his sake. Upshun, too, chinchins Kate again. The people in the Supply are not much pleased with their luck in being sent hither & you, tho' they have arrived as soon as there was much to see, & the Susaquehana goes away when the most interesting part is coming on.

I have no time to write to Dr. Parker, & I think I sent him a line from Lewchew. Give my love to all friends; the knowledge of the heatsome friendship & love of others, to oneself is a pleasant anodyne at all times, & especially when I look to rejoicing them and recording all we have passed tho' while separate. And this joy I hope by the goodness of God is still in store for us.

I cannot expect another opportunity to write you, nor can I at all say when I shall get back, for even the Commodore himself is unable to guess. He wishes to go here & go there, but while he will not go to all, he will doubtless manage to visit many places, and perhaps Ningpo, Fukchan & Amoy may be among them, for I suppose he will not return northward again. All these things are still undecided & doubtful, so that they prevent me saying even a probable time; and you must not remain in Canton longer than you usually do, but, make preparations for a place in

Macao in good time, even if you hire a house. This being obliged to manage for yourself may be troublesome at first, yet I have no great doubt, but that you will soon get accustomed thereto. If you have to hire a house, Mr. Perdon or Mr. Hitchcock will assist you, I think, and Mr. Bonney help to get ready.

I received a number of late papers from Dr. Bridgman by the Supply, which have afforded me & others much pleasure. Pegram, Manry, Mr. Jones & others besides have read them. John is as lazy on board ship as elsewhere; his intimacy with Draper is no benefit. Dr. Morrow is not much liked by the others, which is rather a sign in his favor. I get on pretty well in the Powhatan, which has a diff't mess entirely from the Susaquehana, & better. If I can get a fancy piece of furniture for you, I cannot see where I can take it on board, so full is the ship with Perry's things, tho' I will do what I can.

I send you some Japanese visiting cards; the [an illegible word/ dejuctacies] are quite fanciful about exchanging cards, & no wonder when a man's name is as hard to write & read as [an illegible word/ thesis]. The bosom of the dress of the Japanese is large enough to contain many things, & it is usually occupied with a bundle of papers, writing apparatus, pipe, watch, inkstand, nose-papers, handkerchiefs, purse, medicine bag, & something to eat afterwards, besides two long swords stuck in the belt, in a most uncomfortable manner. I think the Chinese dress is superior for convenience & elegance to this.

Goodbye & sweetest treasure. Keep up good spirits by casting all your cares on One whose heart is large enough to take in all your wants & woes, & power great enough to furnish all you really desire. Soon both of us must go by ourselves to the home whence none return, there to give account of all our conduct to each other. A kiss to the darlings & you from

Yours ever

Wells.

[SWW to his brother Frederick and family: 1854; 03/ 20; 03/ 23; SWWOC]

Yokohama, a village in Japan,
15 miles S. of Yedo,
March 20, 1854

Dear Friends,

I sent you a short line from Lewchew by way of Canton, and now I have begun another here, on this land of mystery, surrounded by dozens of Japanese wondering that ideas be communicated by such unintelligible lines as these, their own being doubly crooked and hard of acquisition. You will be glad to receive a letter from me, I am sure, and to learn that the chief objects of Perry's Expedition to Japan are in good train of completion, and that America and Japan are likely erelong to commence an intercourse. The squadron of 9 ships is now anchored in a [an illegible word/ bight] of this famous & fine bay of Yedo, about 14 miles from the city, and we have the convenience of a house on shore near a poor hamlet of fishermen called Yokohama. I am daily sent ashore to do the talking for both parties, the Japanese who wish explanations of things, & our workmen who are putting up the steam engine, laying the track, displaying the agricultural implements of which there are a large number brought out, and doing whatever else they can to amuse themselves. I have Chinese with me, who amuses himself & the strangers around him, with writing poetry for them, & talking [this] the end of a pen upon this & that. The insulars have never before seen a live Chinese,

& to have him able too to write & communicate with them is an additional pleasure. When I first came here, I made a poor interpreter, having had no practice in this language for nearly ten years, and consequently forgotten the names of almost everything. Here I have plenty of practice, and am likely to get more, if our stay here is prolonged. We get along slowly owning to the want of persons able to intercommunicate; and the Japanese seem to feel the desirability of supplying this deficienty more than we, as they are constantly collecting words; and while I am writing are interrupting me to learn the names of such things as are about them.

The house we are in is a thatched shed, divided into five parallel rooms, having one appropriated to interviews and other purposes of our use, one for exhibiting the rail-road, ploughs, &c.; & the others for the many guards, cooks, & others, officials & attendants, whom the Japanese have around us. All we do & have is as new & strange to them as their dresses, actions, equipages, & doings are novel to us; if each could easily speak to others, intercourse would soon become general.

On our arrival in the Bay, Feb. 13th we found that commissioners had already been appointed by the siogun to negotiate with Com. Perry, the principal of whom was called Hayashi, or Commissioner Lin in Chinese, the 3rd or 4th member of the council at Yedo; his colleagues were Ido, prince of Tsus-sima, Izawa, prince of Mimasaki, & Udono, member of the Revenue Board. There have been several interviews, conducted on their part without the least assumption of [hanteur], or desire to obtain acknowledgements form us of inferiority, and so far as we now know, with frankness and good faith. The progress has been slower than the officers of the squadron have desired, for we are not allowed to go about the country, and peep into everything as we should like; but are most of the time kept on board ship, and live on ship's stores. Yet one, à priori, might expect the commencement of intercourse with such a secluded suspicious people would be slow; and yet has opened more favourably, & I hope will eventuate in a well arranged treaty of commerce. Doubtless, the quiet persuasion of our steamers has had its influence. Blessed be God who has thus far prevented any collision, or even any suspicion of treachery on the part of either party, and who will open the way in his own manner for bringing in the knowledge of his glorious gospel. What extent of trade there will be between the two countries is a matter of much speculation hitherto; and I think mutual wants must first arise before much trade can take place.

The people are, as far as we can see, much cowed down under their rulers, stooping even to the earth when spoken to by an officer, and exhibiting cringing servility equaled only by the haughty air of the superior. Those living in the little fishing-village near us are hearty & fat amidst the meanness of dress & dwelling, as if they had enough to eat, and not much left to mend their condition. The houses are reeds plastered with mud, or thatched walls, or thin clapboards, every sort of being cheap as can be, and roofed with a thick thatch of straw. Domestic animals are exceedingly few; hogs, cattle, sheep, geese & ducks not seen; & poultry, dogs & cats very few. Therefore our chance of getting supplies is not very good, and happily we are not much in want of them. This village & others near it have disappointed our expectations derived from reading their books, but those works on Japan say little of the condition of the common people. The soil is hereabouts a rich black volcanic sand, & the country is divided into small plats of wheat, planted in rows not [broadcast], whose vigor shows the richness of the soil. Great labor must have been bestowed on soil & [crop] to bring them to such a thriving cleanly appearance, and this crop is to be succeeded by rice.

The people are more inquisitive than the Chinese, and not disposed to answer our questions, or else they think a lie will do bother than the truth. Many of them even at this end of the country have learned our letters & can speak a few words of Dutch; the language is more flexible than Chinese, but not so well adapted for foreign sounds as our own.

The Russians are also knocking at the other end of Japan for addmittance, & will probably get much the same admission that we do; afterwards England, France, Peru & Spain will come, until the Japanese will fully know their fellowmen.

[Wells encloses some Japanese greetings cars.]

23rd.

I was in hopes that John would have added a line, telling you a few additional particulars, but he has not done so. His habits have contracted that dilatory, dawdling tendency which life on board ship readily takes & he does almost nothing. His business with the telegraph gives him very little to do, and the residue is not so profitably spent as I wish it was. I send this sheet on to Canton, whence it will go forward, I hope with some additions & inclosures. Love to your both & the babies.

Affectly yours,

S.W.W.

Paid to Alexandria

Rev. W. Frederick Williams

Mosul

Care of Henry A. De Forest M.D.

Beirut

Syria.

S. Wells 1854 Mar. 20, Japan. June 25,

July 17.

[List of Presents: 03/13/1854; 03/ 13; SWWOC]

List of Presents given to the Emperor of Japan and his high ministers by the U.S. Government. March 13th , 1854.

1st. For the Emperor

One ¼ size miniature locomotive, tender, car & track.

A Telegraph appartus, with three miles of wire.

One Francis Life Boat.

One surf boat of copper.

Collection of Agricultural Instruments, as per annexed list.

Audubon's Birds of America, 9 vols.

Natural History of the State of New York, 16 vols.

Annals of Congress, 4 vols.

Laws & Documents of the State of New York.

Journal of the Senate & Assembly of New York.

Lighthouse Reporter, 2 vols.

Bancroft's History of the United States, 4 vols.

Farmer's Guide, 2 vols.

A full series of the United States Coast Survey.

Norris' Engineering.

A dressing-case, Broadcloth.

1 pr. Scarlet Velvet.

A series of U.S. Standard weights, balances, gallon, yard, bushel & gallon.

One ¼ cask of Madeira.

One barrel of Whiskey.

Box of seeds.

One hamper of champagne.

One box of cheery-cordial.

One box of Macarchino.

Three ten-catty Boxes of fine Tea.

Maps of several of the different States.

Four large Lithographs.

One Brass Telescope & stand in box.

One sheet-iron stove.

Six dozen Assorted perfumery, soaps, essences, pomatum &c.

Catalogue of New York State Library.

2 Mail bags with padlock.

Box of arms, containing 5 Hall's Rifles, 3 Maynard's muskets, 12 Cavalry Swords, 6 Artillery Swords, 1 Carbine, & 20 Army Pistols.

2nd. For the Empress

1 Flowered silk Embroidered Dress.

1 Gilded Toilet dressing case.

Six dozen assorted Perfumery, soaps, essences, oils, &c.

3rd. For Abe, Prince of Ishi, 1st Councilor

One copper Life Boat.

Kendall's plates of Mexican War & Ripley's History.

One box of Champagne.

Three boxes fine tea.

One block.

One stove.

One Rifle.

One Sword.

Fifty Gallons Whiskey.

One Revolver & box of Powder & caps.

4 yds. Scarlet Broadcloth.

2 doz. Assorted Perfumery.

4th. For Makino, Prince of Bizen, 2nd Councilor.

Lossing's Field Book of the Revolution, 2 vols.

Cabinet of Natural History of State of New York.

Ten gallons of whiskey.

One dozen Assorted Perfumery.

One clock.

One sword.

One Revolver.

One Rifle.

One large Lithograph.

5th. For Matsudaira, Prince of Izxumi, 3rd Councilor.

Lithograph View of Washington City, with a plan.

One Clock.

One Sword.

One Rifle.

One Revolver.

Ten Galls. Assorted Perfumery.

6th. For Matsudaira, Prince of Iga, 4th Councilor.

Documentary History of State of New York, 2 vols.

Lithograph of a Steamer.

One dozen Assorted Perfumery.

One Clock.

One Rifle.

One Sword.

One Revolver.

Ten galls. of Whiskey.

7th. For Kuzhei, Prince of Yamato, 5th Councilor.

Downyng's Country Houses & book of plan[t]s & fossils.

Lithograph of San Francisco &c.

Ten gallons of Whiskey.

One Clock.

One Rifle.

One Sword.

One Revolver.

Nine sorts of Perfumery.

8th. For Naiïto, Prince of Ki, 6th Councilor.

Owen's Geology of Minnesota & book of plan[t]s. & fossils.

Lithograph of Georgetown, &c.

Ten galls. of Whiskey.

One Clock.

One rifle.

One revolver.

One sword.

Nine sorts of perfumery.

9th. For Hayashi, Dai-gaku no kami, Member of Council and Chief Commissioner.

Audubou's Quadrupeds of America.

One tea set of Whiskey.

Three boxes Fine Tea.

4 yds. Scarlet Broadcloth.

One Clock.

One rifle.

One Sword.

Revolver & box of power & caps.

Box of Champagne.

Two doz. Assorted Perfumery.

One stove.

10th. For Ido, Prince of Tsu-sima, 2nd Commissioner.

Appleton's Dictionary, 2 vols.

Lithograph of New Orleans.

5 galls. of Whiskey.

Nine sorts of perfumery.

Brass Howitzer & carriage.

One Sword.

One Rifle.

One Revolver.

One clock.

One Revolver.

One clock.

Box of Cherry Cordial.

11th. For Izawa, Prince of Mimasaki, 3rd Commissioner.

Model of a Life-boat.

5 galls. Whiskey.

Lithograph of Steamer "Atlanctic".

1 Rifle.

One Clock.

One Sword.

One Revolver.

Box of Cherry Cordial.

9 sorts of Perfumery.

12th. For Udono, Member of Board of Revenue, 4th Commissioner.

List of Port Officers.

Lithograph of an Elephant.

One Rifle.

One revolver.

One Sword.

One clock.

5 galls. of Whiskey.

Box of Cherry Cordial.

9 Assorted perfumery.

13th. For Matsusaki Michitaro, 5th Commissioner.

Lithograph of a steamer.

6 kinds of Perfumery.

One Revolver.

One Clock.

One Sword.

5 galls. Whiskey.

Box of Cherry Cordial.

For Kurokawa Kaheyoye.

14 ps. Striped & white cotton.

Demijohn of Whiskey.

1 Revolver.

4 sorts of perfumery.

Box of Cherry-cordial.

For Moriyama Yenoske, Chief Interpreter

Webster's Dictionary.

Revolver in a box.

Box of cotton goods.

Box of cherry cordial.

6 sorts of Perfumery.

Breaker of Whiskey.

For M Tatsunoske, Second Interpreter

9 ps. Striped & White Cotton.

5 galls. Whiskey.

1 Revolver.

1 Hall's Rifle.

6 kinds of perfumery.

For Namura Gohachiro, Third Interpreter

9 ps. Striped & white cottons.

5 galls. Whiskey.

1 Hall's Rifle.

6 kinds of perfumery.

List of the Present received from the Japanese Commissioners on the 27th March, 1854.

1st. For the United States government from the Emperor of Japan.

1 Gold Lacquered Writing Apparatus.

1 Do Do Paper Box.

1 Do Do Bookcase.

1 Lacquered Writing Table.

1 Pronze Censer on a stand, shaped like an ox with a silver flower on his back.

1 set of two lacquered Trays.

1 Lacquered Bamboo flower holder & stand.

2 Braziers with silver tops.

10 ps. Fine red pongee.

10 ps. White pongee.

5 ps. Figured crape.

5 ps. red dyed flowered crape.

2 Swords & 3 matchlocks.

2nd. For the United State's Government from Hayashi, First Japanese Commissioner

1 Lacquered writing Apparatus.

1 Do Do Paper Box.

1 box of flowered paper.

5 box of stamped note paper.

1 box of flowered note paper.

1 Lacquered Chowchow Box.

1 Box of coral branch and "silver feather" of byssies.

4 boxes of 100 assorted kinds of seashells.

1 box lacquered cups, three in a set.

7 boxes of cups & spoons set from couch shells.

3rd. For the United States Government from Ido, Prince of Tsus-suima, Second Japanese Commissioner

2 boxes lacquered waiters, two in each.

2 boxes 20 paper Umbrellas.

1 box 30 coir brooms.

4th. For the United States' Government from Izawa, Prince of Mimasaki, third Japanese Commissioner

1 ps. Red Pongee.

1 ps. White pongee.

8 boxes of 13 dolls.

1 box bamboo articles.

2 boxes bamboo low tables.

5th. For the United States' Government from Udono, member of Board of Revenue, fourth Commissioner

3 ps. Striped crape.

2 boxes 20 porcelain cups.

1 box soy 20 jars.

6th. For the United States' Government from Matsusaki Michitaro, fifth Japanese Commissioner

3 boxes porcelain cups.

1 box figured matting.

35 bundles oak charcoal.

7th. For the United States' Government from each of the other five Imprerial Councilors, viz. Matsudaira Prince of Iga, Matsudaira Prince of Idzumi, Makino, Prince of Bizen, Naito Prince of Ki & Kuzhei Prince of Yamato

10 ps. of striped & figured pongee or taffeta.

13th. For Commodore Perry, from the Emperor of Japan

1 Lacqured writing Apparatus.

1 Do paper Box.

3 ps. red pongee.

2 ps. flowred crape.

3 ps. dyed figured crape.

14th. For Capt. H.A. Adams from the Japanese Commissioners

3 ps. plain red pongee.

2 ps. red dyed figured crape.

24 lacquered cups & covers.

15th-17th. For Intr. S.W. Williams, O.H. Perry & A. Portman, each from the Japanese Commissioners

2 ps. red ponge.

4 ps. red dyed figured crape.

10 lacquered cups & covers.

18th-22nd. For Messrs. Jesse Gay, R. Danby, J. Morrow,

J.P. Williams & W.B. Draper, each from the Japanese Commissioners

1 ps. dyed crape.

10 lacquered cups & covers.

23rd. For the Squadron from the Emperor

300 fowls.

200 bags of rice, each 160 lbs..

24th. For the U.S.A. From the prefect & Interpreters

5 ps. figured crape.

[SWW to his wife Sarah: 1854/ 06/ 16; 06/ 30; SWWOC]

Simoda, June 16, 1854

My dear Wife,

I take a few moments to drop you a line, but I fail of getting a letter ready when the Southampton sails, for I either have no time to write, or too little strength when I have time, not that I am ill or weak, but I mean that strength of disposition to write you a heartsome letter when I am fagged & sleepy at 10 P.M. I sent you a note from Hakodadi, which I hope is even now on its way from Shanghai from the Vandalia, as she ought to be there by this time, now 17 days. We had a pleasant trip up to that place & back again, and have had warm but good weather since we reached this place. The five commissioners, with two additional ones, forming a respectable row of shaven [pated] gentlemen, have met the Commodore almost daily, since we reached there, to consult upon various regulations pertaining to the trade of the port. You would be much amused to see us all sitting in a little room covered with mats, and partitioned off with folding screens on three sides, the fourth being curtained off from a garden; above is hung a curtain of purple crape, with the coat of arms of the 3d Commissioner, Izawa, and behind the screens & curtains on one side are the dumb gods of the commissioners, sitting in solemn darkness, unable to prevent their temple being made a council house. You would be amused, I say, to see us sitting in this little room, these seven shaved [pated] men on one side, robed in their gowns & big trousers of various colors, each carrying a sword in their girdles like a skewer stuck thro' their midriff, and the Commodore & his four aids on the other side, clad in tight dresses, brass buttons, &c., with cakes & candies & tea before each, on a long bench. The Japanese interpreter sat on the floor, and waddled back & forth as he spoke to one party or the other. At the foot of the room were squatted a dozen Japanese officials of various grades and functions, as still as mice. We formed a remarkable group doubtless, and the import & results of our discussions may prove of great consequence to the Nations of both parties. These deliberations are now brought to a conclusion, and the chance that I now avail of, is while waiting for the commissioners to come aboard to an entertainment prepared for them, both on board of the Powhatan & Mississippi, a dinner & singing by the Ethiopian Minstrels. The latter is dressed up with flags and various embelishments, and presents a very pretty appearance, the light shining thro' the various colored flags. The Japanese are mightily pleased with our music, & the dancing of the Ethiops is diverting in the highest degree. The Commodore had a large bundle of Valentines given him out of the dead letter office, and took them one day to one of our solemn discussions greatly to the amusement of the Japanese, some of the Commissioners laughing till they cried.

The town of Simoda contains about 7000 inhabitants, who live in small one storied houses, and have very few comforts, according to our ideas of comfort. The surrounding country is highly picturesque, presenting a diversity of hills & ravines, valleys and peaks, cultivated, barren, grassy, & wooded, all mixed up and contrasted as Nature knows so well how to do, and man so often looks at as if he had no eyes. The country is not so well peopled as about Yokohama, but if a large commerce gradually grows up here, it is likely that Simoda may by & by become a large town. The officers appointed to live here are going to build a number of fine houses for themselves, & this will involve many artisans, & other laborers, besides the shopkeepers and others connected with trade.

We have found the Japanese on the whole a goodnatured people but not so cleanly as I had read of; in truth, I do not believe a heathen people can be a neat people as a whole nation, for poverty & dirt & ignorance go together, and a cleanly poor man is alone the product of Christianity, much more a cleanly, educated poor man. Women here are degraded & vicious below the Chinese, and thankful ought we to be every day of our lives that we were not born in Japan. You know something how bad the Chinese are, but this people has reached a lower depth than they.

We have now been away from China over five months, and are soon to leave Japan on our return, tho' as I know not how long we are to remain at Lewchew. I cannot say definitely how long it will be ere I can see you and those little folks I so much love & think about. Their playful capers, winsome smiles, broken talk, odd questions, engaging intreaties, querulous complaints, all come up to memory, & I pray our Heavenly Father to keep them, you, & me, till we can together rejoice in his loving kindness to us who are so ungrateful. How much love there is in the human heart if we would search out objects for expending it on.

I have not made out to buy as many things as I wish I c[d]. have done, for all have been disappointed in this respect, & I among them. Some presents will be obtained, and these must satisfy, for it is, after we got away, not what there was to get, but what we have, which must answer. Your table-stand for curios must wait awhile, for nothing of the kind can be had in Japan yet awhile, as the people have not that sort of thing. Lewchew ware forms, as you may suppose, only a small portion of the variety of things offered us for sale, and which is to be sure in the poor shops in a poor village like this; I try to get whatever will illustrate the usages of the people.

I hope you & the darlings are now in Macao, refreshing yourselves with the breezes and walks there, which I doubt not you very much need; where you all have found a lodging-place, is sometimes a matter of speculation to me.

June 30th. [1854] near Lewchew.

I had put away this sheet, as not likely to be sent to you, for instead of the Southsampton going to China, we took her in tow when we left Simoda on the 2nd. last Saturday; but to-day a signal has been made by the Commodore that she leaves us here for Hongkong. John [SWW's younger brother] is enjoying good health, & goes to Formosa as aid to Captain Jones in cracking stones, in the Macedonian; that ship & the Supply left Simoda the same day we did, but they will not reach Killon in Formosa under three weeks, as the SW monsoon is blowing strong, and it is dead to windward, a long beat. D[r]. Morrow is in the Southampton, but whether he will be able to see you I cannot tell. When I shall get to Hongkong is not very certain, but doubtless before or by Aug. 1st. if nothing unforeseen happens. I hope Mr. Bonney will not lose all patience, for I have had no power to return before the rest of the squadron, & I think I have been of some use. Give best love to him & other dear friends in Macao & Canton, whom I have often in remembrance & hope erelong to see you in peace, & praise God together for all the goodness he has shown us all. My health is good, & so is that of the squadron generally, tho' we are all nearly eaten out of all stores. Goodbye, dearest, and keep up good heart, praying that God may bless those dear children he has given us.

In love Your dear,

Husband

[SWW to his brother Frederick: 1854/ 07/13; 1854/ 08/ 19; Lewchew; SWWOC]

Napa, Lewchew, July 13, 1854

U.S.S.F. Powhatan.

Dear Friends,

Your letter of 27th. Dec. was received here after the return of the squadron from Japan on the 1st. inst. and tho' I had to consider you as being again in the midst of those burning heats which are so trying, when I read it, yet I trusted, that the same Gracious Hand which had kept you & me and all those from I heard at the same time, would still keep you & make you the instruments of doing all the good you could do. Since I have read Layard's second work, I am more interested in the diffusion of the Gospel among the tribes & peoples he gives an account of, and hope that the American churches are appointed to lay some of the stones in the living temple to be erected in Mesopotamia.

It is strange to see how we are getting over the world. Already have we set foot in these outer Asiatic nations, and by the Treaty of Kanagawa, the Regulations of Simoda,

& the Compact of Napa, have done much to introduce the Lewchewan, and the Japanese into the com[mun]ity of nations. This port of Napa, & the two in Japan of Simoda & Hakodadi will form doors of entrance, which cannot be shut, even if their inhabitants should desire it. I have been away from Canton seven months, and hope to return there within another, and to find all well.

I have spent a pleasant, & I think useful time. I was not able at first to talk much, and found that the language of these Lewchewans differed so much from Japanese, that it was not worth while to try to acquire its distinctive differences in the expectation of being of any use; while, too, the general acquaintance among the higher classes of Chinese, both spoken & written, obviated all necessity of it. Last year I had little opportunity of talking Japanese, as we were in the country only a fortnight, but this visit afforded more opportunities, and I became generally known at Simoda as the tsuüzhiru no shto, or interpreter, and every urchin would sing out Uriyamusu! whenever I came in sight. The Japanese language does not easily adapt words with many consonants, and some of our names are wider from the original than mine.

Perry has done his mission very creditably, and accomplished as much as any one acquainted with the Japanese had any reason to expect. Business men will doubtless be greatly disappointed in their hope of gain, for the Japanese have opened these two ports to make the profit themselves; and it will be sometime too, before such supplies as ships require can be prepared, & grown. In all ~~Japanese~~ the 3 places the fleet remained at, we saw no sheep, hogs, goats, ~~and~~ or geese; and only a few cattle, chickens, ducks, for fish & products of the sea are the animal viands of the people. Chickens, ducks & eggs were the only land animal products we got of them; but they eat whatever the waters bring forth. In their houses, shops, temples, and general condition, there was not as much comfort, elegance, or indices of taste and industry as I had expected; but I ought not to have expected it, for I had lived among heathens long enough to know what paganism can only produce—the works of the flesh are these: —and the catalogue, with all its attendant dirt, disease, distress, and general degradation, all applies to Japan as it does to all heathendom, alleviated by peculiar makeweights. Dark is the heathen's prospect beyond the grave, & in the doleful, dismal, life he has to lead in his world, he has a foretaste. The Japanese have the elements of rapid progress in arts, knowledge & comforts, in their general inquisitiveness, good nature, and manageable language than the Chinese, but there are drawbacks in the feudal society, with its hereditary privileges and rights, which will make a severe struggle before the rights of conscience are granted.

We shall leave this harbor & beautiful island of Lewchew in a few days more. Dr. Bettleheim leaves his station in charge of a Rev G. H. Moreton, who, with wife & child, are to remain here the only foreigners in the kingdom, and try to live before this people an example of what Christianity can produce. This is about the only efforts he can make, since he is prohibited collecting a school, or dispensing medicine, or entering houses, or preaching the gospel. He enters houses furtively, and talks of the things of the kingdom by stealth, but thus will not the word of God always, be bound. His efforts may seem misdirected & his purposes null, to men of the world, but good gospel seed has God for its husbandman, & will bring forth its fruits in due season, here as elsewhere. The officers & men of the three ships here have given him $275 for the use of the mission, besides other presents of eatables. He has difficulties before him, & a severe trial of patience, yet I look to hear of his fighting a good fight.

I have enjoyed excellent health since I left Canton six mo^{s}. ago, & hope erelong to see the dear one, in as good health as I left them. Peace be with you in all the troubles whh surround you, and prosperity in your soul whatever luck besides the people among whom your lot is cast. John is now in good health, but I dread his course will end in something sad. He loves drink, at least in company, or is easily led away in the midst of others who ridicule those who will not drink. He has gone to Formosa & Manila now, in the Macedonian, & will be in China in Septr. next. I wonder where W^{m} W^{ms}. sons will get to next?

Love & peace be with you
& yours & all the church
in your midst.
Affectly Yours
S. W. Williams

P.S. Canton, Aug. 19/ 54.

I reached here just 7 mo^{s}. from this day I left, & praise God for his goodness in preserving all the dear friends here while I was absent, especially the children & their mother. We have all come together again full of God's goodness. Com. Perry goes home viâ Europe, and will find much honor doubtless from the report of the achievements he has done. Would that he knew how to honor his God, too, who has used him for so distinguished a purpose?

John goes home in the Mississippi, to sail soon. Let us not fail to daily commend him to God's grace that his heart may be touched. He seems to be indifferent to everything, to have no plan or expectation; a weif that floats to every current, obnoxious too to bad, rather than, good influences.

I have a large pile of notes from you, viz. May 6, Dec. 27, April 3, Dec. 31, Feb. 25th & Nov. 19th; of these only one had been seen before I got here. You exceed me quite in letters, but not in love. Your college mate Doolittle is here with his wife; his uvula is so long that he has had it cut off, & is not yet well. He is otherwise well. A delightful work of grace is going on near Amoy, and converts are showing forth fruit to the glory of God. The hope entertained of the good the Taiping wary insurgents would do are rapidly fading into disappointment at their fanaticism & disorders.

Four German missionaries, 3 of them women, have died at Hong Kong since Feb.

[notes in the margins]

S. Wells.

1854 July 13 & Aug. 19 Dec. 30 Nov[em]ber 10.

[J.C. Eldredgh, Purser, to SWW: 1854/ 08/ 30; SWWOC]

U.S. Steam Frigate "Powhatan"

Hong Kong, Aug. 30th, 1854

My dear Sir,

Enclosed I transmit a statement of yours, to Shumming and Awan's account, together with a draft for Two thousand four hundred and five dollars, and twenty four cent, ($2405) being in full payment of them all.

I saw the Comm. P. respecting your acct. after his return, and after talking the subject over, we came to the conclusion, it would be better to allow a liberal time after the return of our ship from Japan, to cover your incidental expense, and loss of time in travelling from and to Canton &c. —so fixed upon the 21st inst., as the period when your appointment would terminate, as you will observe be reference to statement of your account.

To have left anything for reference to Dept. of State,

and to be settled hereafter, would only have complicated matters and given up all additional trouble, so that on the whole, I think the present settlement a proper and liberal one, and hope it will be satisfactory to you. I shall probably see you soon, when we can speak further on the subject if necessary.

In haste
Your Very Briefly
J.C. Eldredgh

[J.C. Eldredgh, Purser, to SWW; Account; Powhatan; 1854/ 08/ 31; SWWOC]

S. Wells Williams, Interpreter to the Japan Expedition, in account with U.S. Steam Frigate Powhatan.
Cr.

By pay as Interpreter, from Feby.19, 1854, on which time you came on board Powhatan to Augt. 31. 54, the date of your detachment 6 months & 12 days, at $300~ per month.

By pay as Rations for 194 days, at 20x for diem.~~~ $ 1920.00~~~38.80.

By pay as Amount due you as transferred from Susquehanna by Purser Barry~~~ [$]1797.91~~~ 3756.71
Dr.

To amount, overpaid by Purser Barry and checked in his account, for pay and rations, from Augt. 9, 1853 to Dec. 31, 1853, 4 months & 22 days, at $300~ pr diem~~~ $ 1419.99.

To Rations for 145 days at 20x diem~~~29.00~~~ $ 1448.99.

To Cash paid by Purser Eldredge including $15~ paid Mr. Lowry for mess a/c~~~196.85.

To Flannel & soup from Pursers Department~~~4.24.

To Hospital fund, 6 mos & 12 days at 20x per mo[nth]~~~1.28~~~1651.36.

$2105.35

Amount due and unpaid, Two Thousand one Hundred and five dollars, and thirty five cents ~

U.S.S. Powhatan

Aug. 31, 1854

J.C. Eldredge

Purser

No. 335. Aman, Landsman, in account with U.S. Steam Frigate Powhatan

Cr.

By pay as Lands from Feby.19, 54, to Aug.7, 1854, inclusive, 6 months, 21 days at $ 9~ per. months~~~$ 54.30.

By pay as Rations for 170 days at 20x for diem~~~34.00.

By pay as Transfer from Purser Barry~~~11.44~~~99.74.

Dr.

To cash paid by Purser Eldredge including Rations to Mr. Lowry~~~$36.40.

To Hospital fund 6 mos & 1 day at 20x~~~1.21~~~37.61.

$62.13

Amount due and unpaid, sixty two dollars and thirteen cents.

No. 49. Loo Shumming, Chinese Interpreter in a/c with U.S. Steam Frigate

Loo Shumming, Chinese Interpreter in a/c with U.S. Steam Frigate Powhatan

Cr.

By pay as ass[istan]t Interpreter, from Feby. 19, 1854 to Aug. 10, 54, inclusive, 1 quarter & 82 days at $ 500 pr. annum~~~$ 236.98.

By pay as Rations for 173 days at 20x per diem~~~34.60.

By pay as Transfer from Purser Barry~~~76.60~~~348.18.

Dr.

To small stores from Pursers Dept.~~~$ .28.

To cash paid by Purser Eldredge, including Rations to Commi. & W. Room messes~~~109.00.

To Hospital fund 5 mos & 21 days at 20x~~~1.14~~~110.42~~~$237.76.

Amount due and unpaid Two Thousand and thirty seven dollars, and seventy six cents~~~~~~~

Reception

Due S. Wells Williams Esq.~~~

$2105.35

Due Aman~~~[$]62.13

Due Loo Shumming~~~[$]237.76

$ 2405.24

[Undated and unidentified document written by SWW: pp. 2; relevant to Lewchew]

The Commander-n-chief of the American squadron now in the harbor of Lewchew hereby makes known to the Regent and other officers of the land his clear determination. For many months ships of his squadron have been anchored at Napa, and their officers and men have uniformly treated the people of Lewchew kindly, have paid for what they have obtained, and made no disturbance. The authorities of Lewchew have not acted friendly, and have not done as they have promised to do; they prevent their people from furnishing supplies, and do not do so themselves; and the people are taught to keep aloof, or to regard the Americans as enemies. Now the Commander in chief has power in his hands, and if he came as an enemy he could use it; but he has restrained it, for he came as a friend, & expects to be treated as a friend; and he herely notifies the Regent that unless the authorities of Lewchew allow the people

to furnish supplies and have free intercourse with the American ships, it will be very easy for him to go to Shui and demand all these things. Let the Regent therefore issue plain orders to the people, and no longer prevent them from buying or selling such things as they have, that no evils may hereafter come upon the authorities or people of Lewchew.
[a scriblling on blank space saying "Kanezaō is a village[;] Natsu Sima Webster I[;] Saru Sima Perry"]

[Commodore Perry to the Regent of LewChew Court; a note, undated and unidentified, but written in the handwriting of SWW; p. 1]
Note.

The course of proceedings of the authorities of LewChew with respect to the mon intercourse, restrictions which they are imposing upon the Officers and [~~men~~] Crews of the US Squadron, restrictions which are uncalled for, and entirely at variance with the terms of friendship which they seem to profess; and considering that all our acts since our first arrival at this Island have indicated the strongest evidence of our peaceful and friendly intentions, I conceive, that I should be acting contrary to reason and common sense, were I to submit any longer to a system of deception, subterfuge and falseholld altogether unnecessary and incompatible with the respect due to the American flag.

I therefore again demand of His Highness the Regent and the authorities of LewChew that these restrictions be removed.
[Another note written by SWW: undated and fragmentary]

It seems strange that persons who have devoted their lives to the preaching and practice of doctrines which they believe were intended for all mankind, and never promulgated by the Son of God himself, who commanded his followers to go throughout the world for this purpose, should have lived so long among [~~the~~] an educated people like the Chinese, and never done moer to teach their adherents the "lively oracles of God". What have been the results in that land of the energetic efforts of the Catholic missionaries, and how small [~~have~~] has been their progress during their existence, in elevating, educating, & converting the people of China, their own publications show. No wonder the father of Marquis (Tseng the present Chinese Minister to England) once tried to remove all apprehensions of danger to China from this source, by showing that the missionaries had obtained their converts by tempting moe to join them through mercenary motives.

The circulation of the Scriptures in Chinese has, however, excited the R.C. Bishops, who have issued their orders to the faithful not to read, keep, or lend [~~the~~] such publications but to burn them immediately. Bishop Besi of Nanking in 1847 [~~issued~~] gave his injunctions on this point warming them in explicit terms "not to take or read the heretical writings which, have issued from Shanghai; & if any adherent of our religion have already accepted them, they must forthwith burn them or deliver them up to their spiritual fathers. These heretical works are of one & the same class with corrupt & obscene writings; and friends of the religion must in no wise either receive them for perusal or hold them in possession". However, as the Chinese Government has [~~never~~]issued no similar proclamation in late years against the distribution of the Bible, [~~the~~] copies [~~of them~~] are constantly coming into [~~their~~] their hands of the people connected with these [~~spiritual fathers~~]churches.

[Commodore M.C. Perry to SWW: 1853/ 03/ 13; New York]

New York March 13th 1853

My Dear Mr. Williams,

I wrote you some time since from Washington upon the subject of the report on narrative of my late Cruise to Japan, which Congress has called upon me to furnish and I naturally look to you, as one conspicuously engaged in the Mission, for valuable aid in the way of notes &c. as also for the Vocabulary to which you should affix your name, —and of which any number of extra copies can be struck off with no farther additional expence than the paper and press work.

Besides the Vocabulary I know of no one who is as capable of writing an historical sketch of Japan as yourself, with the aid of your valuable library and from our personal observations — some forty or fify pages or more might be produced, which would deflect high credit on yourself and furnish a valuable acquisition to my report —

I have no means of renumerating you, but to set aside a certain number of the copies to be assigned to me to be placed at your disposal and of giving you any share of [illegible word/ ultimate] advantage that the copy rights would give to me for the right w^{d}. belong to the Gov^{t}., Congress is generally liberal in giving the [illegible/ affcem] certain privileges.

I shall be impatient until I hear from you and hope you will write by return of mail, and write fully. I should be greatly disappointed not to have your assurance.

My plan is to retain the services of Lieutenants Mausey & Bent to Constant & prepare the charts for the engravers. Messrs Heine & Vallenn to prepare the drawings & Keluber &c for the artists, and Chaplain Jones to superintend the branches of the Nat History assisted by the [one illegible word] of the Smithonian Institute Professor Henry Lavong volunteered such aid.

The truth is every incident connected with the Japanese

expedition, is looked upon with great interest, and there is one universal demonstration of appause of every event which has occurred, and I feel a nervous desire to make the report alike creditable in doing so I depend confidently upon your services and the way I have suggested.

You write with so much ease and so graphically that it will not give you half the trouble it would many others.

Do not forget to send me some translation of Japanese poetry as also Chinese done into English if you have any—these scraps can be apporiately introduced. The Specimans furnished to the Hong Kong Register by your Chinese clerk are quite interesting.

The Susquehannah is in I shall write to Purser Barry to inform me immediately if there should have difficulty respecting the continuance of your salary, during the interval between the two visits to Japan.

Do you think you can send me any materials for a map of Japan, upon which with the data to be obtained from Von Siebold to get up something better than has hitherto been published. This can be done under the law authorizing the publication of my report, and additional copies can be struck off the same as the vocabulary.

My best respects to Mrs. Williams and all inquring friends

Most Truly Yours

M. C. Perry

P.S. Will you in your letter refer to such parts of the Chinese Repository, as [one illegible word] trial of Japan and to such other books, as maybe helpful to me.

[W L Marcy, Dept. of State, to SWW: 1855/ 06/ 28 & 1855/ 09/ 05; pp. 2; "Copy of WL Marcy's letter June 28 1855, & Sept 5, 1855. Reply of SWW, No. 1, 1855"]

Extract

Dep^{t} of State,
Wash^{n} June 28, 1855

The a/c whh the Pres^{t} has rec^{d} of your capacity & qualifications has induced him to app^{t} you Sec. & Int^{r} to the China Mission. This step has been taken without any knowledge or intimation of what may be your wishes upon the subject. The Pres. will be much pleased to learn that the app^{t} will be agreeable to you, & that you are so situate that you can devote proper attention to the affairs of your country in China. Should you be willing to restrain the position, I doubt not you can do so; but should the holding of it be against your wishes, or incompatible with other indispensable engagements, the President is very solicitous that you sh^{d} accept & retain it until he can make another fit selection. Had Dr Parker remained in China, it is probable that he w^{d} have been offered the app^{t} of comm^{n}, & if his health sh^{d} be such as to allow him to return to that country it may be that offer will be made to him. However, on this subject I am not authorized to disclose what will be the course of the Pres^{t}.

I do not propose in this comm^{n} to give you any particular instructions in relation to your duties further than to direct you on receiving & accepting your commission, herewith sent, to take possession of the archives of the Legation & assume charge of the Mission. The instructions given to former Comm^{rs} which you will find on the files of the Legation will be resorted to as a guide to your official conduct, so far as they are applicable. The very great distance of China from the seat of the Gov^{t} of the US, —the long delay in receiving & transmitting comm^{rs} between it & our public agents, —& the constantly changing aspect of affairs there, renders it necessary to leave much to the discretion of the person in charge of the Legation. The Pres[t] reposes full confidence in your judgment & ability, & does not doubt that if you accept the position offered to you, the interests of the U.S. will be in safe hands. Should no Comm[r] or Envoy Ex. be sent out before the time arrives for opening the negotiations for the revision of our treaty you will be furnished with particular instructions on that subject.

If you accept the app[t], your salary which is fixed by law at the vale of $2500 per anm. will commence from the time of our entering upon the discharge of the duties of the office, whh you will advise the Bankers of the US at Canton & likewise this Dep[t].

I am Sir Respectfully
Your ob[t] Servant
W L Marcy

In a 2[d] dispatch (Sept 5/ 55) Mr. Marcy writes:

"I have to inform you that the Pres[t] has appointed Peter Parker Eny Comm[r] of the US to China, & that that gentleman is expected soon to set out on his mission. I make this comm[n] to advise you that it is the wish of the Pres[t] if agreeable to yourself, that you should continue the discharge of the duties of Sec & Chin Int. of the Legation upon the arrival of the new Comm[r], & at the same time to direct you to deliver over to him the books, & archives of Leg of the US at Canton."

Notes

1. Bridgman was the very first American missionary appointed by ABCFM, arriving in Canton on February 19, 1830, and he established a mission press there in 1832, which S. Wells Williams was invited to join one year later. See *Glimpses of Canton, the Diary of Elijah C. Bridgman, 1834~1838*, Yale Univ. School Library Occasional Publications No. 11, 1998.

2. "In 1845, the health of Mrs. Hobson having failed to such an extent, a return to Europe seemed imperative, and he left Hongkong with her in July; but she died within sight of her native land on December 22nd, when anchored off Gungeness, leaving a son and daughter under the care of her widowed husband. During his stay in England, Dr. Hobson was married to the daughter of Dr. Robert Morrison the missionary to China", *Memorials*, p. 125.

3. "Young King had left Brown University and come out as Olyphant's clerk in 1826. He was a very devout Christian and something of a scholar whose humourless piety and outspoken opposition to the opium trade made him intensely disliked, especially among drug traders at Canton. His success as a business man only added to his unpopularity", *The Golden Ghetto*, p. 201.

4. "T.T. Meadows had fallen in love with the language as instantly and irrationally as others fall in love with a woman at first sight. After attending s single lecture at Munich University by a professor who strangely combined the teaching of Armenian and of Chinese, Meadows abandoned all studies other than Chinese, took himself out to China intent on a government post and, after a brief spell

as an assistant, was found competent to be an interpreter", *The China Consuls*, p. 12.

5. "As the ships entered, the British ensign was seen suddenly to rise on a flag-staff, placed near a house, which was perched on a curious overhanging point of rock, north of the town; this house was the residence of the missionary, Mr. Bettelheim, a convert from Judaism, who married in England, and had, for some years, been resident on the island, under the auspices of an association of pious English gentlemen, officers of the British navy, very much, however, against the inclination of the Lew Chewans", *Commodore Perry and the Opening of Japan*, p. 146.

6. "The telegraph apparatus, under the direction of Messrs Draper and Williams, was soon in working order." Hawks, Commodore Perry and the Opening of Japan, p. 342. "Williams" in the quotation is John P. Williams (1826~1857), one of the younger brothers of SWW.

7. "Commodore Perry being desirous of obtaining as full information as possible of Peel Island during his short visit, determined to send parties of exploration into the interior. He accordingly detailed certain officers and men for the purpose, who were divided into two companies, one of which was headed by Mr. Bayard Taylor, and the other by Dr. Fahs, assistant surgeon", *Commodore Perry and the Opening of Japan, op. cit.*, p. 204.

8. "Though the Flora of Japan does not come strictly within the province of our investigation, we thought it might not be out of place to give here a list of the novelties discovered in that country (in connection with Dr. Morrow) by D. Wells Williams, the great sinologist, whose merits with respect to the Botany of China we shall have to record in another chapter. The figures refer to the pages of Asa Gray's paper. 306. *Clematis Williamsii*, A. Gray. Simoda,

April 20th, 1854.—Maxim, Mél. IX (1876) 601....,” *European Botanical Discoveries in China*, Vol. 1, p. 389.

9. A personal note on Mr. Timothy E. Head from the East West Center, and his interest in the history of the Bonin Islands. When he reached Tokyo more than 45 years ago, he had already got a very interesting and charming book published, *Going Native in Hawaii* (Charles Tuttle, Tokyo), based on his own experiences visiting one island after another in Hawaii, with his wife besides him, and with an indispensable cooking equipment, called *shichirin* (a small-sized charcoal brazier) on his back. He was one of my dear college teachers, and he took me many times to the National Library in order to read Japanese books on the Bonins. It was a hard training for me to translate them into English almost instantly, for he was patient but ready to make notes. I hope he has written as charming a book, on the life of Savary, as the former one.

10. On June 3rd, 1834, Parker embarked at New York in the *Morrison* for China, which same ship carried him with SWW, King, and his wife, Gutzlaff and seven ship-wrecked Japanese sailors to Japan in 1837. He wrote about his experience in *Journal of an Expedition from Singapore to Japan with a visit to Loo-choo; descriptive of these islands and their inhabitants; in an attempt with the aid of natives educated in England, to create an opening for missionary labours in Japan*, published in London, 1838.

11. J.H. Levyssohn: *Bladen over Japan: Met eene afbeelding van Decima*, S Gravenhage, 1852.

12. “Hibiscus Rosa sinensis. Sp.II, 977. H. foliis ovatis acuminatis serratis, glabris, caule arboreo. Variat flore pleno. Habitat in India. “Breyn, Centur. exot. 121, t. 56. Alcea javanica arborescenes, flore pleno rubicundo”, *European Botanical Discoveries in China, op. cit.,* p. 72.

13. "S. Wells Williams, a lay-missionary who had been twenty years on the China coast as a managing editor of the Mission Press for the American Board of Foreign Missions. He did not profess to know either the Okinawan dialects or Japanese, but he was a distinguished scholar of the Chinese language, fully competent to supervise diplomatic correspondence in that language. His knowledge of the Ryukyu was extensive; he had carefully studied the Chinese records of Okinawa's position in the Chinese tributary system; he had visited Naha in the 'Morrison' in 1837 and had published an important account of that experience. As editor of the *Chinese Repository* at Canton he had seen masses of Bettleheim's correspondence and has published some of the missionary's letters in 1850" ,Kerr, *op. cit*., pp. 306-307.

14. "From a commanding height above the ships, the view is, in all directions, picturesque and delightful. On one hand are seen the distant islands, rising from a wide expanse of ocean, whilst the clearness of the water enables the eye to trace all the coral reefs, which protect the anchorage immediately below… the eye being, in every direction, charmed by the varied hues of the luxuriant foliage around their habitations." McLeod: *Voyage of His Majesty's Ship Alceste, along the Coast of Corea, to the Island of Lewchew*, 2nd Edition, John Murray, London, 1818, pp. 101-102.

15. "The interpreter of the regent was a young native, *Ichirazichi*, who had been educated in Pekin, where he remained three years. He could speak a little English, but the Chinese was the language of communication. This youth had some knowledge both of the United States history and geography. He was not unacquainted with the character and conduct of Washington, and called him 'a very great mandarin'", *Commodore Perry and the Opening of Japan*,

op. cit., p. 194.

Shimadzu Nariakira, 島津斉彬 , Prince of the Satsuma clan "rewarded Itarashiki Satonushi (the 'Ichirazichi' of Perry account) for his skill in handling the unpredictable and stubborn strangers", Kerr, *op. cit.*, p. 345.

16. Vries Island or Idzu O-sima was found by a Dutch navigaor, de Vries, in 1643, and hence the English name appearing in European charts.

17. "Dr. Williams, of Canton, was present as interpreter of the Japanese language; although his services were not called into requisition. Mr. A.L.C. Portman, the commodore's clerk, as it was most agreeable to the Japanese, acted as interpreter in the Dutch language", Spalding, p. 169.

18. Commodore Biddle anchored "Columbus" here on July 20, 1846. "The only effort we had made toward opening friendly relations, (and it scarcely deserves the name,) was in sending two ships under Commodore Biddle, which remained at anchor some eight or ten days, accomplished nothing, and quietly left when the Japanese desired it", Hawks (1856), p. 77.

19. "It is due to Lieutenant Bent, an officer on board the Mississippi, to acknowledge that the Commodore availed himself of that gentleman's former experiences in a visit in the Preble to pilot the ships as they entered Napha", Hawks(1856), p.150.

20. "In February of the year 1849 the United States ship Preble, under Commander Glynn, formed part of the American squadron in the China seas, when information was received, by way of Batavia, of the detention and imprisonment, in Japan, of sixteen American seamen, who had been shipwrecked on the coast of some of the Japanese islands. The Preble was immediately dispatched to demand

their release. As the ship neared the coast of Japan, signal guns were fired from the prominent headlands to give warning of the approach of a strange vessel; and when she entered the harbor of Nagasaki, she was met by a number of large boats which ordered her off, and indeed attempted to oppose further ingress. But the ship steadily standing on with a firm breeze soon broke their ranks, and came to anchor in a desirable position. " Hawks (1856), p. 48.

21. "On October 8, 1804, Rezanov approached Nagasaki, but though he presented the permit Laxman had brought back, he was not admitted to the port until December 24, 1804, and even then only upon surrendering Russian ammunition for 'safekeeping'." *Russia's Japan Expedition of 1852 to 1855*, p. xxi.

22. "The Japanese whom we had brought with us were then presented to the Great Man, dressed in their silk clothes of Russian manufacture ; and each shewed the silver watch and the twenty ducats with which he had been presented by the Russian monarch. The interpreters begged us to instruct them in the Russian language, and offered to instruct us without any expense in the Japanese. We found the people of distinction here uniformly polite and courteous in their manners: but for their language and costume, we might have supposed ourselves among the most polished Europeans." Langdorff, Part 1, p. 241.

23. S. Wells Williams: *Tonic Dictionary of the Chinese Language in the Canton Dialect*, 1856.

24. The Chinese name is 羅森 (1821-1899), whose Japanese experiences then are compiled into the three instalments of his article, titled 日本日記 , published in "Chinese Seriel" by the Anglo-Chinese College in Hongkong, 1853. Refer to the writings of 老師王晓秋 of 北京大学 who specializes in the study of the first modern

intelligent's visit from China to Japan.

25. John Johnson: American Baptist Missionary Union. He arrived at Hongkong with Mrs. Johnson on January 5th, 1848, and left for a short visit to U.S. in 1858.

26. "In 1393 a Chinese immigrant community of clerks and craftsmen was settled in Okinawa as a direction of the imperial government. The records speak of the 'Thirty-six Families', but this must not be taken as a literal numeration; it was customary to speak of the 'Thirty-six Families of Fukien' or of the 'Hundred Names' of China in figures of speech which merely meant a widely representative group." Kerr: *Okinawa*, p. 75.

27. The Voages and Adventures of Ferdinand Mendes Pinto, the Portuguese.

28. Isaac Titsingh(1744-1812): Nipon o daï itsi ran; ou, Annales des empereurs du Japon.

29. Julia von Klaproth (1783-1835): *San kokf tsou ran to sets*. 2 vls.

30. John P. Williams, one of the younger brothers of SWW. John joined the Perry Expedition as a telegraph engineer, and though he liked to stay on in China, he soon died in the SWW's house in 1857. The fine-grained granite memorial which was put into the headstone of his grave at the East India Company Cemetery in Macau reads as follows: IN MEMORIAM/ JOHN P. WILLIAMS/ OF UTICA/ STATE OF NEW YORK, U.S.A./ DIED AT MACAO/ JULY 25, 1857,/ AGED 31 YEARS. / HE ASSISTED IN SETTING UP/ THE FIRST/ MAGNETIC TELEGRPHA/ IN JAPAN IN 1854.

31. Bretschneider: *History of Botanical Discoveries in China,* vol. 1, p. 387; "On p. 217 Perry states that Mr. W. Heine, who accompanied the expedition as an artist attributed chiefly to the procurement of birds. Botanical

specimens were gathered by the chief interpreter Dr. Wells Williams and Doctors Green, Fahs, and Morrow. On p. 306-332 Professor Asa Gray published an *Account of the Botanical Specimens—List of Dried Plants Collected in Japan (1854) by Dr. Wells Williams, and Dr. James Morrow.* These collections were chiefly made at Simoda, Yokohama, Hakodade".

32. S. Wells Williams, Chief Interpreter.

33. John P. Williams, one of the two telegraph engineers.

34. "Clematis Williamsii. A. Gray. Simoda, April 20th 1854 — Maxim. Mél. IX (1876) 601." See Bretschneider, vol. I, p. 389.

35. See "Yoshida Shonin" in Robert Louis Stevenson's *Familiar Studies of Men and Books.*

36. Spalding, pp. 288-9; "A few days afterward, some of our officers in their strolls ashore, ascertained that there were two Japanese confined in a cage at a little barrack back of the town, and on going there they were found to be the persons who had paid the midnight visit to our ships, and they also proved to be my unfortunate friends of the letter. They did not appear greatly down-cast by their situation, and one of them wrote in his native character on a piece of board, and passed through the bars of his cage, to one of our surgeons present, what follows: "When a hero fails in his purpose, his acts are then regarded as those of a villain and robber. In public have we been seized and pinioned, and darkly imprisoned for many days; the village elders and headmen treat us disdainfully,their oppressions being grievous indeed; therefore looking up while yet we have nothing wherewith to reproach ourselves, it must now be seen whether a hero will prove himself to be one indeed. Regarding the liberty of going through the sixty states (of

Japan) as not enough for our desires, we wished to make the circuit of the five great continents; this was our heart's wish for a long time. Suddenly our plans are defeated, and we find ourselves in a halfsized house, where eating, resting, sitting, and sleeping, are difficult, nor can we find our exit from this place. Weeping we seem as fools, laughing as rogues—alas ! for us, silent we can only be. Isagi Kooda, Kwansuchi Manji."

37. This is the first printing using English movable types in the modern Japan, though done on board a ship. See the entry to June 16th.

38. "Stone House Museum. Belchertown Historical Association. Elijah Coleman Bridgman Papers, 1820-1839, 60 items. *Background note*: Elijah Coleman Bridgman (1801-1861) and his wife, Eliza Jane Gillette Bridgman, spent most of their lives in China as missionaries. After graduating from Amherst College and Andover Theological Seminary, he arrived as a missionary in China in 1829 and lived there as an educator, translator, and publisher until 1861. A native of Belchertown, Bridgman was editor of the Journal of the *North China Branch of the Royal Asiatic Society* and spent later years of his life revising a Chinese version of the Bible. He was prominent in the Medical Missionary Society and the Morrison Education Society, editor of the *Chinese Repository* for the entire length of its publication (1832-1851)." Xiaoxin Wu, ed. by: *Christianity in China* (2nd edition, 2009), p.p. 192-193.

39. "In 1808 an English frigate, H.M.S. *Phœton*, sailed into Nagasaki in search of Dutch prize and threatened to bombard the shipping in the harbour if food was not supplied. The Governor gave orders that the *Phœton* should be attacked and burned, but nothing was done. The intruder, having obtained food and fuel, sailed out again.

This incident, following shortly upon the Russian attacks of 1806, made a strong and disagreeable impression. It caused the Bakufu to take disciplinary steps against the officers at Nagasaki (who had shown a poor spirit), to strengthen the coast defences, and with some prompting from Dutch, to conclude that England and Russia has designs upon Japan." Sansom: *The Western World and Japan,* p. 246.

40. Frederick's footnote says that "The famous monument to Washington at the American capital was not completed until long after Perry's death. The stones collected in Japan, Loo-choo and China are built into its side; the bell, in accordance with the Commodore's wish, was presented in 1858 to the Naval Academy at Annapolis, where it still hangs in an orientalesque frame near one end of Lovers' Lane. I bearts an inscription in Chinese telling of its origin". Frederick W. Williams, edited: *A Journal of the Perry Expedition to Japan 1853-1854,* p. 246.

41. Henry Van Vleck Rankin (1825-1858) was sent by as a missionary to China by the Board of Foreign Missions of the Presbyterian Church, arriving at Ningpo in August, 1849.

42. Daniel Jerome Mcgowan was sent to China as a medical missionary by the American Baptist Board of Foreign Missions.

43. J. A. T. Meadows. "Born Northumberland, Married a Chinese. After resignation became a merchant at Tientsin, where he acted as US vice-consul and consul for various other powers." P.D. Coats: *The China Consuls,* p. 497.

44. McCarthee Family Papers, Presbyterian Historical Society, Philadelphia. "Divie Bethune McCartee (1820-1900) was appointed by the Board of Foreign Missions as a medical missionary to China in 1843. He went to Ningpo in 1844, engaging in medical and evangelical work. In

1853, he married fellow missionary Juana M. Knight. McCartee performed consular services in Shanghai until a regular consular service was established there in 1857. The McCartees returned to Ningpo in 1865, then were transferred to the Shanghai mission in 1872. McCartee resigned shortly thereafter to join the Shanghai consular staff as interpreter and assessor in the Mixed Court." Xiaoxin Wu, ed.: *Christianity in China*, p. 419.

45. Richard Quarterman Way was sent to China as a missionary by the Board of Foreign Missions of the Presbyterian Church, arriving at Ningpo on November 6th, 1844. He had a charge of a missionary Boys' Boarding School there from 1845 to 1853.

46. Josiah Goddard (1813-1854) was sent to China a missionary by the American Baptist Board of Foreign Missions in 1838.

47. Robert Henry Cobbard was sent to China as a missionary by the Church of England Missionary Society, arriving with Mrs. Cobbard at Ningpo on May 13, 1848.

48. The wife of Seneca Cummings (1817-1856), a missionary sent to China by ABCFM, in company with Dr. and Mrs. James, Miss Pohlman, the Rev. CC. and Mrs. Baldwin, and the Rev. Richards, landing at Hongkong on March 25th, 1848.

49. Caleb Cook Baldwin was sent to China as a missionary by ABCFM, arriving at Fuhchow on May 7, 1848.

50. John Stronach was born in Edinburgh on March 7, 1810. He was sent to China as a missionary by the London Missionary Society, reaching Amoy with his wife on July 8th, 1844.

51. William C. Burns was a missionary sent to China by the English Presbyterians. "Although Burns, with all his

devotion and ability, had never been content to stay long in one place, the foundations had been well enough laid by him to be built upon successfully by his successors. Those in and around Amoy patiently nourished the Christian communities in the effort to make them as soon as possible, self-governing, self-supporting, and self-propagating." Kenneth Scott Latourette: *A History of Christian Missions in China* (reprinted by Ch'eng-Wen Publishing Company, Taipei, 1975). p. 247.

52. Elihu Doty was a missionary to China sent by ABCFM. "Elihu Doty, an ordained minister of the Dutch Reformed Church in the United States was married to Clarissa D. Ackley of Litchfield, with whom he left New York, in the beginning of June, 1836." *Memorials of Protestant Missionaries to the Chinese*, p. 97.

53. Mrs. Williams Jones Boone, whose maiden name was Sarah Ameha De Saussure, of South Carolina. She died of the prevailing fever on 30th of August, 1842, in Amoy.

54. Mrs. Elihu Doty died on October 5th, 1845, in Amoy, leaving two daughters.

55. Mrs. William John Pohlman, whose maiden name was Theodosia R. Scudder, the sister of Dr. Scudder the missionary to India. She died on September 30, 1844, in Amoy, leaving three children including an infant daughter of nine days.

56. Mrs. James H. Young, whose maiden name was Sarah Harriet. She died on December 3rd, 1853.

57. John Lloyd. "After consulation with the brethren of his mission there, he left for Amoy with Dr. Hepburn about the end of November, and reached his destination on December 6. There he remained occupied with his missionary pursuits till the time of his death, which took place December 6th, 1848, exactly four years after his

arrival." *Memorials of Protestant Missionaries to the Chinese,* p. 144.

58. One year before in March, 1853, Parkes had written to Mrs. Lockhart telling her about his busy life as temporary Consul in Canton: I am now in charge here (he told Mrs. Lockhart), as Dr. Bowring has gone on three months' leave of absence; but with only assistant, and with an establishment of five reduced to two, we have our hands full. In my new capacity I shall have the honour of marrying the first English couple that (will) have been married in Canton. I must introduce you to the parties: Miss Augusta Fischer and Mr. John Williams, tea-taster in Messrs Jardine Matheson & Co… " I would have written you a respectable letter by this mail, had it not been the old American Commodore Perry, who has visited Canton, and is slowly eating his way through a phalanx of dinners, one of which it fell to my lot to give yesterday, twelve persons at table... If there is anything I cordially detest it is having a 'tall feed', as my United States friends term it. I strive very hard to dine nine days out of ten by myself off a chop or steak, the deglutition of which takes from ten to fifteen minutes…" Stanley Lane-Poole: *The Life of Sir Harry Parkes*, Vol. 1, pp. 179-180.

59. James H. Young.

60. SWW Okinawa Correspondence [SWWOC]: Yale University Library; Williams Family Papers; Correspondence and related papers 1853-1854, in English and Chinese, of Samuel Wells Williams, missionary, diplomat, and sinologue, concerning his service as interpreter on Commodore Matthew C. Perry's expedition to Japan. 96 frames, ca. 28 items; Microfilmed at Yale University Library, September 1970.

61. Gideon Nye: *The Memorable Year — or the War in China; the Mutiny in India; the Opening-up of the*

Resources of Siam; the Projected Movement upon Cochin-China; and the Monetary Crisis in Europe and America; — being a Record of Periodical Reflections and Comments Elicited by the Course of Events in the East, with Incidental Notices of Political and Commercial Affairs in the West and Some Special Papars upon Political and Geographical Topics of the Period; and Included a Sketch of the Inflation and Collapse of Mr. High-Commissioner Yeh. First printed in Macao, 1858, and re-printed by Elibron Classics, 2005.